PRINCIPLES AND PRACTICE

Michael R. Emmert-Buck
John W. Gillespie
Rodrigo F. Chuaqui (Eds.)

Dissecting the Molecular Anatomy of Tissue

With 43 Figures, 9 in Color, and 8 Tables

Springer

Professor Michael R. Emmert-Buck
Dr. John W. Gillespie
Dr. Rodrigo F. Chuaqui

Pathogenetics Unit
Advanced Technology Center
Laboratory of Pathology and Urologic Oncology Branch
Center for Cancer Research
National Cancer Institute
8717 Grovemont Circle
Bethesda, MD 20892-4605
USA

Cover Illustration: Breast cancer. Intraductal carcinoma showing intense membranous staining for c-erbB-2; original magnification ×400. S-ABC technique with hematoxylin counterstaining

ISBN 3-540-22006-2 Springer-Verlag Berlin Heidelberg New York

Library of Congress Control Number: 2004108877

Springer is a part of Springer Science+Business Media
springeronline.com

Printed in Germany

Production: PRO EDIT GmbH, Heidelberg
Cover-Design: Design & Production, Heidelberg
Typesetting: K + V Fotosatz, Beerfelden

39/3150 beu-göh – 5 4 3 2 1 0 – Printed on acid-free paper

Preface

The study of tissues, particularly those derived from the clinic, is a complex and challenging undertaking. Historically, these investigations were morphology-based and descriptive in nature. Although they provided little molecular-level information about the normal and pathological events that occur in specimens, they were essential in determining cellular constituents and the phenotypic alterations that occur in disease processes. In the last few decades, the application of laboratory techniques, such as immunohistochemistry and Northern blotting, have generated molecular results from tissue samples that complement morphological information. These data permit investigators to move toward a mechanistic understanding of histopathology, and are also important in generating clinically useful diagnostic and prognostic markers. In parallel, fundamental mechanisms of pathophysiology are now under study in many laboratories using in vitro (co)-culture and mouse model systems. Yet our understanding of tissues at the molecular level remains woefully incomplete. In particular, basic information on the "-omic" state of each cellular phenotype, including mRNA and protein measurements (in the context of a specific genomic background), as well as the status of biochemical pathways is not available. Using astronomy as an analogy, one might say we are like early observers of the night sky using rudimentary telescopes that measure only the brightest objects and observe a limited spectrum of the electromagnetic radiation in the universe. Clearly, there is much more to know.

The good news is that it is an exciting time for investigators interested in analyzing tissues. The intersection of multiple biomedical disciplines is opening up new opportunities and possibilities. The information from the Human Genome Project provides a "parts list" of the genes that need to be measured. High-throughput technologies, such as expression microarrays, allow simultaneous quantitation of large numbers of transcripts, in the order of tens of thousands. The more recent emphasis on proteomics promises to deliver new methodologies that permit both whole proteome measurements and determination of the activation state of each protein. Finally, new computational techniques developed by bioinformaticians permit the overall data sets to be analyzed in ways that transcend the traditional "one lab–one gene" approach. Ultimately, the field is moving toward histomathematical descriptions of normal and diseased tissue samples, in other words, comprehensive numerical representations of cellular phenotypes. The generation, analysis, and interpretation of these datasets will significantly improve our understanding of cellular pathophysiology, and provide signposts towards new and more effective clinical interventions.

The present book presents both a theoretical and a practical guide to analyzing tissue samples and is divided into four sections; clinical issues, pathology laboratory, molecular analysis, and protocols. The chapters in the first three sections describe available methods, some of the challenges involved with these efforts, and provide examples where new technology(ies) and methodologies may be useful. The last section of the book provides a series of "ready-to-use" laboratory protocols. The authors and topics were selected from a broad range of clinical and scientific disciplines to emphasize the multidisciplinary aspect of this undertaking, especially when clinical tissue specimens are analyzed. No one aspect of the project can be viewed in isolation, thus it is critical to utilize an integrated approach. In that regard, the first section of the book is dedicated to clinical issues. Dr. Moskaluk begins by discussing the historical development of anatomic pathology, from basic science investigation to practical use in the clinic to the current development of high-throughput molecular analyses that complement morphologic examination. His chapter concludes with a description of how molecular analysis of microdissected samples may be clinically utilized in the pathology laboratory of the future. The second chapter by Drs. Kim and Ornstein illustrates the potential near- and long-term patient-care utility of tissue profiling in more detail.

The second section of the book follows samples from the clinic to the pathology laboratory. The chapter by Drs. Gillespie and Gannot describes the various steps in tissue processing and how these may impact subsequent molecular studies. Several accompanying methods are provided in the section on protocols. Dr. Stephen Bova's chapter continues with a discussion of data management in the context of studying large cohorts of tissue specimens. Included is a description of a novel laboratory tool (LabMatrix) that permits investigators to efficiently control, manage and access all of the components that comprise a study. The chapter by Dr. Michael Tangrea and colleagues covers the historical development of tissue microdissection, describes the newer commercial laser-based technologies that are available, and also presents two new technologies for tissue analysis, expression microdissection (xMD) and layered expression scanning (LES), techniques that when fully developed may complement methods currently in use. Dr. Stephen Hewitt, Director of the Tissue Array Research Program (TARP) at the National Cancer Institute, presents an overview of tissue arrays and their utility for studying clinical or animal specimens in a high-throughput manner.

The third section of the book moves from the pathology laboratory to the research bench. Once a specimen has been processed and subsequently subjected to microdissection, there are multiple molecular analysis techniques that can be utilized, depending on the goals of the study. In addition, the upstream nature of the sample (clinical, animal model?) and the effects of tissue processing (snap frozen, fixed and embedded?) need to be factored into both the study design and subsequent interpretation of data. Investigators have a diverse set of available tools, but it is critically important they are used properly to maximize the information that can be gleaned from the specimens, and to ensure that the results are accurate. In Chapter 7, Dr. Sergio González presents traditional techniques that are used for studying tissues, the strengths and weaknesses of the approaches, and how these methods can be used most efficiently in the laboratory. The chapter by Drs. Uribe and Wistuba focuses specifically on genomic DNA analysis of microdissected samples, discussing a range of methods, from genotyping to mutation analysis. Drs. Ozsolak and Luzzi from Dr. Mark Watson's laboratory present a superb overview of RNA analysis, including high-throughput expression arrays and subsequent validation with RT-PCR. Finally, Dr. David Geho of the NCI-FDA Clincial Proteomics Group, co-chaired by Drs. Petricoin and Liotta, rounds out the molecular analysis section by discussing the new and rapidly developing field of "tissue proteomics."

Our hope is that the present book is useful to a broad range of investigators and clinicians who are interested in the analysis of tissue samples. The ability to accurately and comprehensively analyze molecular anatomy has the potential to significantly enhance our understanding of normal physiology and disease processes.

The Editors

Contents

Contributors

Bova, G. S.
Pelican Laboratory, Johns Hopkins University, Department of Pathology, Carnegie Bldg. Room 628, 600 North Wolfe Street, Baltimore, Maryland 21287-6417, USA

Chuaqui, R. F.
National Cancer Institute, Bethesda, Maryland 20892, USA
Pathogenetics Unit, Laboratory of Pathology, NCI/NIH, Advanced Technology Center, Rm. 109D, 8717 Grovemont Circle, Gaithersburg, Maryland 20877, USA

Emmert-Buck, M. R.
National Cancer Institute, Bethesda, Maryland 20892, USA
Pathogenetics Unit, Laboratory of Pathology, NCI/NIH, Advanced Technology Center, Rm. 109D, 8717 Grovemont Circle, Gaithersburg, Maryland 20877, USA

Flaig, M. J.
Klinik und Poliklinik für Dermatologie und Allergologie der Ludwig-Maximilians-Universität München, Frauenlobstr. 9–11, 80337 München, Germany

Gannot, G.
National Cancer Institute, Bethesda, Maryland 20892 USA and Pathogenetics Unit, Laboratory of Pathology, NCI/NIH, Advanced Technology Center, Rm. 109A, 8717 Grovemont Circle, Gaithersburg, Maryland 20877, USA

GEHO, D.H.
FDA-NCI Clinical Proteomics Program, Laboratory of Pathology, Center for Cancer Research, National Cancer Institute, Bethesda, Maryland, USA

GILLESPIE, J.W.
SAIC-Frederick, Inc., National Cancer Institute at Frederick, Frederick, Maryland 21702-1201, USA

GONZÁLEZ, S.
Department of Pathology, School of Medicine, Catholic University of Chile, Lira 85, 4° Piso, Santiago, Chile

HEWITT, S.M.
Tissue Array Research Program (TARP Lab), Laboratory of Pathology, Center for Cancer Research, National Cancer Institute, National Institutes of Health, Bethesda, Maryland, 20892-4605, USA

KIM, I.Y.
Department of Urology, University of California, Irvine, Building 55, Rm 304, 101 The City Drive, Orange, California 92868, USA

KNEZEVIC, V.
20/20 Gene Systems, Inc., Rockville, Maryland 20850, USA

LIOTTA, L.A.
FDA-NCI Clinical Proteomics Program, Laboratory of Pathology, Center for Cancer Research, National Cancer Institute, Bethesda, Maryland, USA

LUZZI V.
Department of Pathology, University of Texas Southwestern School of Medicine, Dallas, Texas 75390, USA

MOSKALUK, C.A.
Department of Pathology and Biochemistry and Molecular Genetics, University of Virginia, Charlottesville, Virginia 22908, USA

ORNSTEIN, D.K.
Department of Urology, University of California, Irvine, Building 55, Rm 304, 101 The City Drive, Orange, California 92868, USA

Ozsolak, F.
Department of Pathology and Immunology,
Washington University School of Medicine,
St. Louis, Missouri 63110, USA

Petricoin, E. F.
FDA-NCI Clinical Proteomics Program, Center for Biologics Evaluation and Research, Food and Drug Administration, Bethesda, Maryland 20892, USA

Tangrea, M. A.
National Cancer Institute, Bethesda, Maryland 20892, USA
Pathogenetics Unit, Laboratory of Pathology, NCI/NIH,
Advanced Technology Center, Rm. 109D, 8717 Grovemont Circle,
Gaithersburg, Maryland 20877, USA

Uribe, P.
Department of Anatomic Pathology,
Pontificia Universidad Catolica de Chile, Lira 85, 4° Piso, Santiago, Chile

Watson, M. A.
Department of Pathology and Immunology/Box 8118,
Washington University School of Medicine, 660 S. Euclid Ave.,
St. Louis, Missouri 63110, USA

Wistuba, I. I.
Department of Pathology and Thoracic/Head
and Neck Medical Oncology, MD Anderson Cancer Center,
Box 85, 1515 Holcombe Blvd., Houston Texas 77030, USA

I Clinical Issues and Tissue Processing

1 Molecular Pathology Laboratory of the Future

Christopher A. Moskaluk

1.1 The Past

The integration of laboratory analysis with human medicine has traditionally been divided into two broad specialties: anatomic/surgical pathology and clinical pathology. Clinical pathology has been primarily the realm of "wet lab" analysis of bodily fluids and secretions, and over time has utilized the full range of laboratory techniques in chemistry, biochemistry, microbiology and molecular biology. The analysis of whole human tissue has been primarily the providence of anatomic pathology, which is much older than laboratory-based medicine, and is a direct descendent of the discipline of gross dissection and examination of human organs of Galen and his contemporaries in ancient Greece. Gross dissection allowed for the basic understanding of the "plumbing" of the human body, and the subdivision of the organs into organ systems of distinct function (circulatory vs. digestive, etc.). During the Renaissance in Italy, autopsies became standard practice in the investigation of human disease. Diseases began to be categorized by the gross changes seen in tissue (e.g., hemorrhage, tumorous growths, caseating necrosis, sclerosis). It is of interest to note that most of these ancient terms have not disappeared, but have been incorporated and integrated into the modern lexicon of clinical medicine, even as the very molecular basis of these diseases is being understood.

With the advent of microscopy in the seventeenth century and the discovery of the cellular basis of life, diseases in which distinctive tissue changes occurred could be categorized on their histologic appearance. Advances in microscopic optics and histologic techniques correlated with increased understanding of disease processes. It was discovered that various substances "fixed" tissues in a way that prevented autolytic breakdown and decay. The development of mechanical microtomes to section tissue as thinly as possible allowed for greater penetration of light through the tissue and provided uniformity of the sections. Embedding of tissue in wax allowed for a rigid matrix that greatly aided in histologic sectioning, but this

Principles and Practice
Emmert-Buck et al. (Eds.):
Dissecting the Molecular Anatomy of Tissue

embedding required the extraction of the water from tissue, hence the stepwise processing of tissue through a graded series of alcohol baths into an organic solvent was devised. Since most tissue elements are essentially colorless, many varieties and combinations of stains have been developed to enhance histologic contrast among the various constituents of tissue architecture. By the latter half of the nineteenth century, the basic histologic technology that is still most widely used today had been developed: the hematoxylin and eosin stained formalin-fixed paraffin-embedded tissue section (Fechner 1997).

Histologic analysis of tissue came into its own as a clinical diagnostic test in the nineteenth century, with the German physician-scientist Rudolph Virchow being its greatest early proponent. It was at this time that the "humoral theory" of human disease (disease being the result of imbalances of specific fluids in the human body) began its rapid decline, with the understanding of the cellular basis of human disease, including infectious and oncologic disorders. It is interesting to note that early academic anatomic pathologists were most interested in the scientific discovery of the basis of human disease and were not especially interested in its direct application to clinical medicine. Especially in the United States it was surgeons, many outside academic medical centers, who were the early practitioners of microscopic tissue examination in the clinical management of patients (Rosai 1997). These clinicians needed methods to determine whether surgery was indicated, what type of surgery was indicated, and if the surgical procedure was successful in its intent. This shift in focus from scientific inquiry to clinical practice is reflected in the terminology of "surgical pathology" to describe the subspecialty of anatomic pathology concerned with clinical tissue diagnosis. This very practical need for better ways to inform clinicians on the selection of therapeutic options drove the ascendancy of histologic tissue assessment. This fact is useful to consider when one ponders the future of new technologies in the application of tissue diagnoses.

1.2 The Present

The majority of tissue-based diagnostics in our time is based on the interpretation of the morphologic characteristics of fixed and stained cells and tissues as seen under the microscope. That this technology of over a 100 years of age is still extensively used speaks for its power. New disease entities are still being defined using these methods, and the histologic classification and categorization of diseases are continually revised as our understanding of, and our experience with, these disease processes increases. Although some disease classification requires the assessment of cell ultrastructure by electron microscopy, this requirement occurs in a small minority of clinical tissue examinations. While the strength of histologic analysis is that tissue and cell morphology is often an accurate reflection of the sum of many molecular alterations, its weakness is that these underlying molecular alterations are not specifically identified.

In general, histologic tissue assessment is concerned with the identification of specific cell types, and the overall organization of the tissue structure. Abnormalities in tissue organization may identify a congenital abnormality, a degenerative process, or an acquired destructive process. Of the latter category, the major etiologies are autoimmune, infectious and neoplastic. Neoplastic processes are usually identified by specific changes that occur on the cell structure level, most commonly an increase in the size of the nucleus, irregularity of the nuclear contours, and/or an increase in DNA content as reflected by uptake of the hematoxylin stain. Architectural changes (abnormal arrangement or placement of cell populations) are also frequently employed in the identification of neoplasms. Once the diagnosis of a neoplasm is established, subclassification is based on differentiation patterns as expressed in cell shape (spindle vs. polygonal, etc.), cell products (mucin, keratin, etc.) and architectural arrangement (formation of glandular structures, single cell infiltration, etc.). The variation from sample to sample due to differences in fixation and staining, as well as the subtlety of many of the pathologic changes has precluded major inroads by computer image analysis, and the interpretation of tissue histology remains the realm of physicians with specialty training in surgical pathology.

There has been some integration of molecular assays into surgical pathology. By far, the most prevalent form of molecular assay is the detection of specific antigens (usually a protein species) by an antibody in the technique of immunohistochemistry. The rapid assimilation of this technique in clinical practice is in part a consequence that the interpretation of this technique is microscopic examination, hence was easily assimilated into the normal practice of surgical pathology. The majority of immunohistochemical analyses employed are for the *classification* of disease processes rather than the determination of *prognosis* of a disease process. For instance, a metastatic poorly differentiated neoplasm may have a panel of antibodies applied to determine if the neoplasm has characteristics of epithelial cells, stromal cells, melanocytes or lymphocytes, hence may be classified, respectively, as a carcinoma, sarcoma, melanoma or lymphoma, with consequences for the selection of subsequent chemotherapy and radiation therapy. While appropriate classification and staging of a clinical disease is important for prognosis, surprisingly there has been little impetus for widespread implementation of molecular tests whose results yield solely prognostic information.

A case in point is clinical testing for her-2/neu overexpression in breast cancer. The discovery that amplification of the her-2/neu oncogene was correlated with a more aggressive cancer phenotype and a poorer prognosis was first discovered in 1987 (Slamon et al. 1987), without the significant adoption of this assay in clinical assessment for over a decade. However, widespread clinical testing for her-2 neu overexpression did not occur until the advent of targeted therapy with trastuzumab (Herceptin) in the late 1990s (Goldenberg 1999; Hanna et al. 1999). The clinical need to assign a therapy has been the driving force in the adoption of a new technique in tissue diagnosis for over a century, and will probably continue to be so for the future.

Other examples of the integration of molecular assays in tissue diagnosis is in situ hybridization, which is slowly being implemented as a standard test to study specific genetic changes that occur in cancer. The indications include subclassification of lymphomas and sarcomas on the basis of specific chromosomal translocations and the analysis of her2/neu gene amplification. Again, note that these determinations usually have implications for the type of anti-neoplastic therapy employed. Polymerase chain reaction (PCR) of genomic DNA or coupled with reverse transcription of messenger RNA (RT-PCR) to identify specific gene rearrangements from tissue samples is also being increasingly used, but remains primarily a technique performed at relatively few specialized laboratories.

1.3 The Future

Leaving the realm of validated clinical tests, there are several technologies currently being used to analyze tissue in the research setting. The most exciting are those that provide a "global" analysis of hundreds to tens of thousands of analytes. This includes DNA microarray analysis, which can be used to assay DNA and RNA content (Duggan et al. 1999; Lipshutz et al. 1999). With a current capacity of tens of thousands of probe sets, the whole of the predicted human gene population can be assayed for the presence and quantity of messenger RNA transcript. Similarly, the probes may query single nucleotide polymorphisms (SNPs) that can determine the presence of specific polymorphic sequences on chromosomes. Such a microarray can be used to determine the gain or loss of specific chromosomal regions in neoplasms. Alternately, microarrays can be used to query all of the nucleotides of a specific gene, allowing detection of all possible mutations in a single assay. For proteins, mass spectroscopy is the most promising global assessment technology. In this technique, protein extracts are ionized, and the constituent proteins are separated in an electric field on the basis of mass and charge. Depending on the specific technique used, either a complex "fingerprint" of the constituent proteins is obtained that may be distinctive for a specific disease process or tumor type, or the actual identity of a subset of the proteins can be determined. All of these technologies impart a catalogue of biomolecules present in a tissue sample, revealing characteristics that may or may not be reflected in the morphologic characteristics of the cells, characteristics that correlate with functional biochemical and biological activity.

The greatest challenge to new technologies is to match the sensitivity of standard histopathology. An experienced surgical pathologist may confidently render an unequivocal diagnosis of a malignant neoplasm with as little as ten cells (or less) from a biopsy that contains thousands or tens of thousands of cells. This sensitivity ranges across most types of the hundreds of cancer subtypes. While some molecular assays are as or more sensitive (PCR for example), none can combine the universality of histologic analysis with such sensitivity. In addition, some tests, especially global analyses of biomolecules, will require a pure population of cells to avoid confounding signals coming from an impure cell population. Given that cancer cells are always present in varying quantities in clinical samples (biopsies, fine-needle aspirates, resection specimens), it is quite likely that a selection or purification step will be required to obtain the cell population of interest. Again, since it will not be known *a priori* what cell type will be present in an initial diagnostic biopsy, histologic analysis will most likely still be employed to make an initial assessment of the disease process.

There have been several studies reported in the literature to show that microarray gene expression profiling can classify cancers as well as traditional histologic examination (Golub et al. 1999; Giordano et al. 2001; Ramaswamy et al. 2001; Su et al. 2001). Are these findings the harbingers of the decline of histology as the major tissue diagnostic modality? Perhaps not. The key fact to remember here is that these genomic analyses were generated from cell populations that were recognized and at least partially purified by histologic examination. Only if molecular markers are identified that robustly and specifically identify discrete cell populations will the need for histologic assessment disappear. The fact that this approach could be used is made clear by the current use of flow cytometry. In this technique, cells in suspension can be identified and separated on the basis of antibody binding, size and cytoplasmic substructure. While flow cytometry on whole cells may be used with ease in noncohesive "liquid" cancers of leukemia and lymphoma, its use in the separation of malignant cells from solid tumors (bound by physical bonds to neighboring cells) is problematic. The manipulations required to dissociate cohesive tissue has major impacts on protein and RNA transcript populations and the selection of separation techniques often requires some *a priori* knowledge of the desired cell population.

The last point brings up the issue of a possible "tissue Heisenberg uncertainty principle" at play. The techniques and procedures that allow assessment of one particular molecular species (e.g., immunohistochemistry, in situ hybridization) often cause changes in other molecular constituents (e.g., RNA and protein degradation, protein dephosphorylation). If it is true that a complex analysis of a large number of specific molecular markers is required to identify a specific cell population, it is difficult to see, therefore, how the need for "simple" histologic examination will be replaced, at least as an initial clinical assessment. Only if two major technical problems can be overcome will be need for histologic assessment disappear: (1) the lack of an adequate *simple* panel of molecular markers that separates out desired target cells and (2) the elimination of major changes in other molecular constituents of the cells during the initial "cell identification" assays.

If histologic analysis cannot be eliminated, and if a subsequent assay yields no or only nominally additional clinically relevant data, economics will dictate that the secondary assay will probably not be adopted in patient care. Does this mean that most molecular analytic techniques will remain solely in the realm of research laboratories, and not move into the clinical arena of tissue diagnosis? Probably not, for it is clear that histologic assessment cannot determine with precision all of the biological heterogeneity inherent to disease processes, especially neoplasms. Reports are emerging showing that such global gene expression surveys can identify subclasses of tumors within traditional cancer classification schemes that respond differently to standard therapy (Pusztai et al. 2003). As discussed above, it is this enhanced ability to ascribe a specific therapeutic regimen that will drive the adoption of these new molecular-based assays in tissue diagnostics.

It would appear that neither will histology be completely supplanted by novel technologies, nor will the need for molecular assessment of clinical tissue abate. Given that histologic analysis will continue, it will probably be most efficient to combine histologic examination with the cell purification step required for the global molecular analyses. The microscopic-guided isolation of specific cell population from tissue sections is termed "microdissection", and several microdissection techniques have been developed to achieve this task (Emmert-Buck et al. 2000). As discussed elsewhere in this volume, the current state of the art tissue microdissection platforms employ laser beams that capture cell populations specifically targeted by the operator by direct microscopic examination. While being almost exclusively research apparatuses today, it is expected that these platforms will continue to develop in optical clarity, ease of use, and automation as they move from research laboratories to the desks of surgical pathologists.

Given that the trend in tissue diagnostics is to render diagnoses on smaller and smaller samples obtained by increasingly minimally invasive procedures (needle biopsies and fine-needle aspiration), molecular assessments will have to be extremely sensitive. The techniques will have to be refined, such that reproducible and robust genomic/proteomic signatures can be generated from a handful of cells. Polymerase chain reaction (for DNA and RNA) and in situ transcription-based RNA amplification technologies now exist that meet or probably soon will meet these requirements, with proteomic technologies requiring further enhancements.

Additional advances that will probably be reflected in the clinical tissue assessment of the future are robotics and digital image analysis. While automated laboratory testing is the norm in clinical pathology, surgical pathology is primarily a bastion of "hand-crafted" skills, dependent on the dexterity of histotechnologists and the pattern recognition skills of the physician. Even today's cutting-edge technologies of laser-assisted microdissection require relative painstaking and time-consuming tissue acquisition by a skilled histologist. In the future, more of the sample preparation and microdissection and the majority of the downstream analyses will be performed automatically on robotic platforms. This will also allow for remote manipulation of the machinery, so the pathologist may not have to be in the same room, or even the same building as the tissue that he or she is analyzing. Image analysis will also probably improve significantly, with some level of computer-guided recognition of target cells within tissue sections. If molecular probes can successfully be employed, then target cell acquisition may be almost completely automated. It is expected that such automated image analysis will always require a final inspection and adjustments by a physician, but having such an initial target selection will greatly increase throughput. Finally, having the images digitally recorded will create a permanent record of the morphologic features, even for specimens that are subsequently destroyed during microdissection and molecular analyses. Thus, in the future, tissue samples may be placed into a "black box" and never again be touched by human hands, but their essence on both a morphologic and molecular level will be determined and digitally recorded as a permanent virtual tissue sample (Fig. 1.1).

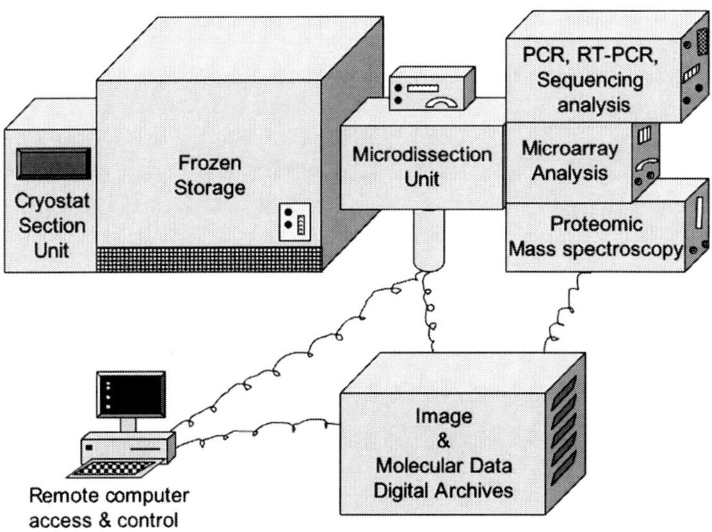

Fig. 1.1. Surgical pathology workstation of the future. Tissue samples are fed into the cryostat sectioning unit at *left*. Histologic sections are made, examined and microdissected robotically, controlled remotely by technicians and physicians. Microdissected tissue samples are processed for various specific and global molecular assessments. Images and molecular data are stored digitally for analysis and archiving

1.4 A Scenario of the Future of Tissue Diagnostics: A Prostate Biopsy for Assessment of Carcinoma

1.4.1 Current Clinical Procedures

Let us first walk through the steps that typically occur today in a needle biopsy assessment of a man suspected to have prostate cancer. After the biopsies have been obtained by sampling the organ with a hollow needle, the physician places the tissue in a buffered formalin solution that forms chemical cross-linking bonds between the macromolecules. The tissue is transported to the pathology laboratory, where it is then placed in a machine that automatically changes the solution surrounding the tissue to increasingly high levels of ethanol, slowly replacing the water present in the tissue. The ethanol is then replaced by an organic xylene solution at elevated temperatures. Finally, the organic solution is replaced with molten paraffin that permeates the substance of the tissue now that the aqueous content has been removed. This "tissue processing" step typically is an overnight process. The next morning, a technician removes the tissue from the processing unit, and transfers it to a mold containing molten paraffin. The paraffin is allowed to harden, encasing the tissue in a rigid matrix, called a paraffin block. The paraffin block is transferred to a microtome, where a technician cuts thin sections from the block and mounts them on slides. The tissue on the slides are subsequently re-hydrated, generally a reverse of the tissue processing procedure, except that it only takes minutes given the thinness of the section. The tissue section is then stained and cover-slipped. The slides are delivered to the surgical pathologist with the accompanying paperwork documenting the identity of the tissue specimen. The pathologist examines the section under a microscope, and decides to order an ancillary immunohistochemical stain, which takes another day to complete. The next day, after examining the stain, he dictates his findings, which are transcribed into a report that looks something like this:

- Gross Description
  Received in formalin are five 0.1 mm diameter needle biopsies of tan tissue ranging in length from 0.5 to 2 cm. The tissue is submitted in its entirety for histologic examination.
- Microscope Examination
  Tissue consists of prostate with varying degrees of chronic prostatitis and glandular atrophy. In one core, comprising approximately 5% of total cross-sectional area, are a few well-formed glands of epithelium with neoplastic features, infiltrating around normal structures. An immunohistochemical stain for keratin 903 shows a lack of basal cells within these glandular elements, confirming the histologic impression of an invasive carcinoma. Perineural invasion is not identified.

- Final Diagnosis
  Prostate, needle biopsies: prostatic adenocarcinoma, Gleason score $3+3=6/10$.

This hypothetical case shows in essence the extent of tissue diagnosis for the majority of cancer diagnoses in the present day: the presence or absence of cancer, the histologic classification of cancer (in this case prostatic adenocarcinoma), the grade of the cancer (in this case a scheme known as Gleason scoring), the presence or absence of other known histologic predictors of tumor behavior (in this case perineural invasion). Occasionally, as in this case, immunohistochemistry is used to confirm the presence of a neoplasm, or to help in the subclassification of a neoplasm.

1.4.2 Future Clinical Procedures

Now let us look into the future and walk through the steps that might accompany a similar specimen. After the biopsies have been obtained, they are dropped into a container with an embedding gel that contains substances to stabilize biomolecules, including RNA and protein phosphorylation sites. The container is immediately dropped into liquid nitrogen (which is now present in all clinics throughout the hospital), and the specimen is transported to the pathology laboratory in a frozen state. A technician takes the tissue (embedded in the frozen gel) and places it in an automated cryostat that cuts thin tissue sections onto many glass slides, and then robotically bar codes them, and moves them to a –140 °C freezer for intermediate storage. One of the middle sections is automatically stained and cover-slipped, then moved to an integral microscopy workstation.

The surgical pathologist who will be responsible for the sample analysis, has decided to work from home this day. His laptop (outfitted with a high definition screen) gently beeps on his kitchen table. He looks up from his coffee, and establishes a secure network connection to the hospital laboratory. He begins to analyze the cases in his queue for the morning. A real time histologic image of the prostate biopsy appears on his screen. The pathologist controls the movement of the slide in the x- and y-axes, as well as the power of the objective and the focus controls remotely. The frozen section is not of the same histologic quality as a standard histologic section, but there is an area architecturally suspicious for carcinoma in one small area of one of the needle core biopsies. A few clicks of the computer mouse, and the image analysis software confirms the presence of enlarged nuclei and nucleoli in this cell population, consistent with carcinoma. Satisfied with this assessment, the pathologist makes a few more mouse clicks, then moves on to the next case. In the meantime, three of the frozen slides are immediately robotically retrieved, stained, and moved to a microdissection station.

When the pathologist is finished with the day's initial histologic assessments, he is ready for microdissection. Still operating the machinery remotely from his computer, he calls up the prostate biopsy case and with the help of automated image analysis software, quickly locates the suspected area on each section and using a selection tool draws circles on the screen over the location of the malignant glands, identifying the target tissue for the apparatus (Fig. 1.2, color illustration at end of book). A few more mouse clicks and the assays are selected and the microdissection apparatus begins automated tissue capture and transfer to the analysis machinery. The pathologist has selected microarray analysis of global RNA transcript expression, a mass spectroscopy analysis of phosphorylated protein residues, and mutational analysis of six genes that are standard for prostate cancer analysis. Any and all extraction, purification, amplification and analytic techniques are carried out robotically at that point, without human intervention. All of the analyses are carried out in parallel and are completed overnight. The next morning the pathologist reviews the accumulated data and releases the final report. The gross and microscopic de-

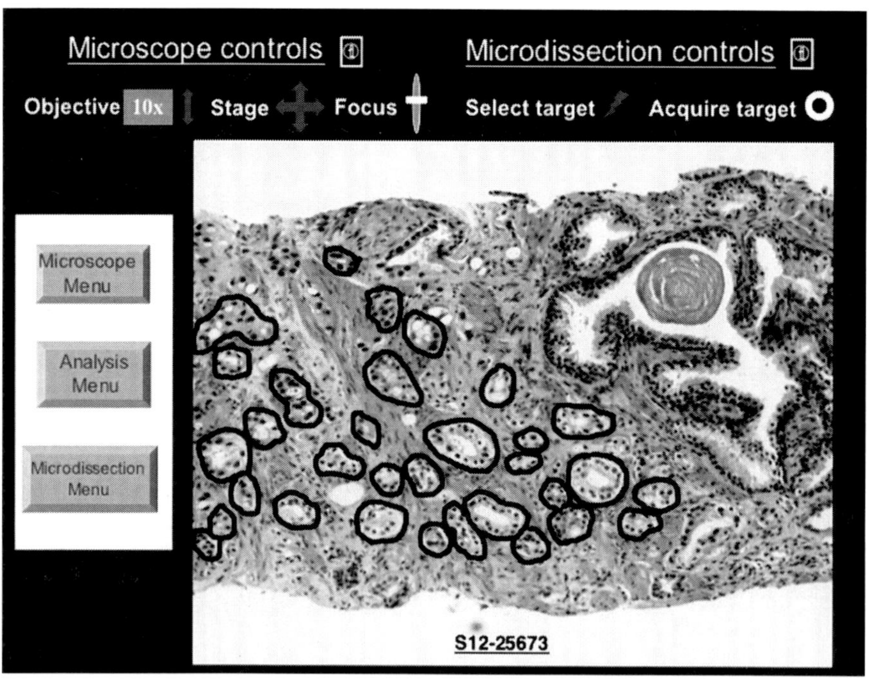

Fig. 1.2. Future scenario for prostate cancer diagnosis on biopsy specimen. On a computer screen, a pure population of prostate cancer cells is selected on a real-time image of a histologic section prior to automated microdissection and molecular analysis (see color illustration at the end of the book)

scriptions remain essentially the same as the above example. The final diagnosis is a bit more detailed:

- Histologic Diagnosis
  - Prostatic adenocarcinoma, Gleason score 3+3=6/10.
- DNA Analysis
  - P53: wild type, no deletions
  - PTEN: wild type, no deletions
  - PRCA6: frame shift at codon 45 (inactivating), deletion of second allele
  - AR: wild type, 12× gene amplification
  - Pros2: Asn123His (activating mutation)
  - TCFNG: wild type, no amplification

  Gene mutation summary: risk of metastasis moderate
- RNA Transcript Profile
  - Subgroup PRCA-R-C3b (gene-specific data on Lab Medicine Genomics Server, with predicted chemotherapy responses)
    - Risk of metastasis: moderate–high
    - Predicted sensitivity to ionizing radiation: 54th percentile
    - Predicted risk of androgen insensitivity: 73rd percentile
- Phosphoproteome Profile
  - Subgroup PRCA-PPh-12 (gene product-specific data on Lab Medicine Genomics Server, with predicted chemotherapy responses)
    - Risk of metastasis: moderate–high
    - Predicted sensitivity to ionizing radiation: 63rd percentile
    - Predicted risk of androgen insensitivity: 80th percentile
- Clinical Recommendations
  - Radical prostatectomy (if patient is surgical candidate), with pelvic lymph node sampling.
  - Adjuvant therapy (protocol # PCPRC324), followed by dual androgen blockade with agents PAR24a and duplicenterol.

In this prediction of the future of clinical tissue diagnosis, several assumptions are made. The first is that "global molecular assessment" technologies have advanced in sensitivity to be applied to a few tens of cells. The second assumption is that the heterogeneity of tissue constituents continues to be a major confounding variable in molecular analysis, and solution-based assays, microarray, and mass spectroscopy applications will require target cell purification by microdissection. The third assumption is that advances in therapeutics have kept apace with molecular diagnostics, so that targeted therapies are present that will require the molecular assessment of tissue. It is likely that in the future, as it is today, the majority of such tissue assessments will be for the diagnosis of neoplasia, with assignment of cancer subtypes for the purposes of treatment.

1.5 Summary

Tissues are complex collections of different cell types. Diseases affect the number, type and function of these constituent cells. The major diagnostic modality used today to diagnose disease processes in tissues is histologic examination, a technique that relies primarily in recognizing changes in the morphology of the tissue and constituent cells. Because of the shifting paradigm of modern medicine, in which increasingly the molecular basis of human disease is targeted with specific therapies, tissue diagnostics will need to keep apace. In addition to assays that measure one analyte at a time, global genomic and proteomic assessments will also likely be integrated into clinical practice. The complex nature of tissue will likely require the integration of traditional histologic examination and microdissection with multimodality molecular analyses that will increasingly integrate robotic platforms, and digital image analysis. Physicians, presumably surgical pathologists, will need to master new skills to carry out and interpret the analyses required to fulfill the promise of molecular-based therapeutics.

References

Duggan DJ, Bittner M, Chen Y, Meltzer P, Trent JM (1999) Expression profiling using cDNA microarrays. Nature Genetics 21(1 Suppl):10–14

Emmert-Buck MR, Strausberg RL, Krizman DB, Bonaldo MF, Bonner RF, Bostwick DG, Brown MR, Buetow KH, Chuaqui RF, Cole KA, Duray PH, Englert CR, Gillespie JW, Greenhut S, Grouse L, Hillier LW, Katz KS, Klausner RD, Kuznetzov V, Lash AE, Lennon G, Linehan WM, Liotta LA, Marra MA, Munson PJ, Ornstein DK (2000) Molecular profiling of clinical tissue specimens : feasibility and applications. Am J Pathol 156:1109–1115

Fechner RE (1997). The birth and evolution of American surgical pathology. In: Rosai J (ed) Guiding the surgeon's hand; the history of American surgical pathology. American Registry of Pathology. Washington, DC, pp 7–21

Giordano TJ, Shedden KA, Schwartz DR, Kuick R, Taylor JM, Lee N, Misek DE, Greenson JK, Kardia SL, Beer DG, Rennert G, Cho KR, Gruber SB, Fearon ER, Hanash S (2001) Organ-specific molecular classification of primary lung, colon, and ovarian adenocarcinomas using gene expression profiles. Am J Pathol 159(4):1231–1238

Goldenberg MM (1999) Trastuzumab, a recombinant DNA-derived humanized monoclonal antibody, a novel agent for the treatment of metastatic breast cancer. Clin Ther 21(2):309–318

Golub TR, Slonim DK, Tamayo P, Huard C, Gaasenbeek M, Mesirov JP, Coller H, Loh ML, Downing JR, Caligiuri MA, Bloomfield CD, Lander ES (1999) Molecular classification of cancer: class discovery and class prediction by gene expression monitoring. Science 286(5439):531–537

Hanna W, Kahn HJ, Trudeau M (1999) Evaluation of HER-2/neu (erbB-2) status in breast cancer: from bench to bedside. Modern Pathol 12(8):827–834

Lipshutz RJ, Fodor SP, Gingeras TR, Lockhart DJ (1999) High density synthetic oligonucleotide arrays. Nat Genet 21(1 Suppl):20–24

Pusztai L, Ayers M, Stec J, Hortobagyi GN (2003) Clinical application of cDNA microarrays in oncology. Oncologist 8(3):252–258

Ramaswamy S, Tamayo P, Rifkin R, Mukherjee S, Yeang CH, Angelo M, Ladd C, Reich M, Latulippe E, Mesirov JP, Poggio T, Gerald W, Loda M, Lander ES, Golub TR (2001) Multiclass cancer diagnosis using tumor gene expression signatures. Proc Natl Acad Sci USA 98(26):15149–15154

Rosai J (1997) Some considerations on the origin, evolution, and outlook of American surgical pathology. In: Rosai J (ed) Guiding the surgeon's hand; the history of American surgical pathology. American Registry of Pathology. Washington, DC, pp 1–5

Slamon DJ, Clark GM, Wong SG, Levin WJ, Ullrich A, McGuire WL (1987) Human breast cancer: correlation of relapse and survival with amplification of the HER-2/neu oncogene. Science 235:177–182

Su AI, Welsh JB, Sapinoso LM, Kern SG, Dimitrov P, Lapp H, Schultz PG, Powell SM, Moskaluk CA, Frierson HF Jr., Hampton GM (2001) Molecular classification of human carcinomas by use of gene expression signatures. Cancer Res 61(20):7388–7393

2 Clinical Utilities of Microdissected Samples

Issac Yi Kim and David K. Ornstein

2.1 Introduction

Tissue microdissection is a developing technique that is gaining a strong foothold in the biomedical community. For example, since the initial report on laser capture microdissection (LCM) by Emmert-Buck et al. in 1996, this highly innovative technology that allows the procurement of pure populations of cells and recovery of macromolecules for molecular analysis has been fully embraced by researchers. Molecular changes relevant to human disease can now be studied in human tissues, obviating the need to rely on cell lines for investigations that depend on pure populations of cells. Information generated from microdissection-based studies is already being used extensively in experimental analyses in the laboratory, as well as in the design of novel clinical diagnostic tools and therapeutic strategies. However, in contrast, the application of this technology into patient care has been comparatively slow. Nevertheless, there have been multiple reports that demonstrate the potential usefulness of tissue dissection and it is likely that the method will become an important tool in the future armamentarium of the surgical pathologist. To date, tissue microdissection has been used to improve diagnostic accuracy of cytology preps, detect viral DNA, determine gene copy number, and measure cell cycle parameters. In this chapter, these potential clinical applications are reviewed with a focus on the use of LCM.

2.2 Current Clinical Uses

2.2.1 Diagnosis

Laser capture microdissection can isolate a group of cells from complex tissues, such as prostate, containing a mixture of normal, pre-malignant, and malignant areas without damaging the morphology of the selected and

Principles and Practice
Emmert-Buck et al. (Eds.):
Dissecting the Molecular Anatomy of Tissue

surrounding zone. Based on these early reports, numbers of laboratories have begun to utilize LCM to enhance sensitivity and specificity of diagnosis; this improved diagnostic capability, in turn, influences follow-up regimens and treatments rendered to the patients. For example, Paradis et al. (2000) used LCM to identify cirrhotic patients with an accumulation of monoclonal nodules. Because hepato-cellular carcinoma (HCC) has a very poor prognosis and most HCC arise from malignant transformation of regenerative cirrhotic nodules, detection of premalignant lesions may improve the management of patients with cirrhosis. To identify those patients at risk for malignant transformation, the authors utilized LCM-dissected nodules for clonal analysis because monoclonality is a hallmark of neoplasia. Using this approach, the investigators were able to identify 57 of 112 micronodules that were monoclonal and recommended that these patients be followed carefully.

2.2.2 Prognosis

The cornerstone of tissue diagnosis is histologic changes at the cellular level; both morphologic and architectural changes are used for diagnostic and prognostic information. However, in some cases an accurate histologic diagnosis may be difficult to make because the observed differences are subtle and buried within a field of normal cells. In this situation, molecular analysis of LCM-procured cells holds promise to improve the prognostic information beyond what can be achieved with standard histopathology. Traditionally, tumors are graded based on architectural and cytologic features. These grading systems usually provide powerful prognostic information for low- and high-grade tumors, but frequently provide no useful information for intermediate-grade tumors (which often represent the majority of patients). For example, in prostate cancer, Cooperberg et al. (2003) recently surveyed the Cancer of the Prostate Strategic Urologic Research Endeavor (CaPSURE) database and reported that intermediate-grade tumors comprised 75.4% of prostate cancer diagnoses. Since tissues are a heterogeneous mixture of different cell types, and pertinent molecular alterations are often specific to a single cell type, isolating a pure population of cells is critical to obtain clinically relevant molecular information. Multiple studies have shown that profiles produced from cleanly dissected samples are superior to those from bulk tissue specimens, and likely will be important in future efforts to determine prognosis based on molecular features of disease processes. For example, Miura et al. (2002) showed that by cDNA microarray analysis, cells which were laser capture microdissected from lung adenocarcinoma cases can distinguish smokers from nonsmokers and more importantly, distinguish between those patients who survived less than 5 years from those who survived longer than 5 years after surgery.

2.2.3 Cytology

Cytology can be problematic because often there are a small number of malignant cells in the presence of a large number of background cells. Recently, it has been demonstrated that LCM can be used to isolate a homogeneous cell population from a cytologic specimen that can be subsequently used for further molecular analysis. In this report, the investigators analyzed ten cases of paraffin-embedded tissues and cytologic specimens from the same patients using LCM to isolate a homogeneous population of B-cell lymphoma cells; genomic DNA was then isolated from the captured cells (Orba et al. 2003). Subsequent PCR analysis for immunoglobulin heavy chain (IgH) gene rearrangement confirmed the diagnosis of B-cell lymphoma. The authors reported that DNA was easily obtained from Papanicolaou-stained or immunostained specimens that were preserved for 1–3 years. However, air-dried specimens that have been Giemsa stained were not amenable to LCM because cells were firmly attached to the glass slides. In addition, the authors suggested that the cytologic analysis appeared to have a higher sensitivity than histologic analysis in diagnosing B-cell lymphoma. The authors proposed that the reason for the observed enhanced sensitivity in diagnosing B-cell lymphoma with LCM is due to the collection of targeted cells homogeneously with minimal contamination and to the absence of immediate ethanol fixation. In short, LCM can be used to enhance the diagnostic accuracy of cytology and should facilitate molecular analysis of limited cytologic specimens.

2.2.4 Detection of Viral DNA

Although viruses are an important cause of many diseases, including some cancers, it can be difficult to detect viral particles from suspected tissues because only a few cells may be infected. LCM can facilitate detection of viral particles. For example, investigators were able to detect Epstein-Barr virus (EBV) in cells isolated by LCM in a study examining the link between EBV and hepatocellular carcinoma (HCC) in Japanese patients. The connection between EBV and HCC was initially made based on the following observations. First, monoclonal EBV DNA was detected by Southern blot analysis in 13 of 35 Japanese patients of HCC tumors; 10 of these cases contained EBV only in tumor tissue and not in nontumor tissues (Sugawara et al. 1999). Second, EBV was detected by PCR in HCC from patients positive for hepatitis C and B virus (HCV and HBV, respectively); though EBV was present more frequently in the HCV group (Sugawara et al. 2000). However, for tumors to be designated as "EBV-associated", Junying et al. (2003) felt that an unequivocal demonstration of the EBV genome for viral gene products in the tumor cell population is necessary; in this respect, both PCR and Southern blot are unsatisfactory since they cannot accurately localize the virus to specific cell types. To overcome the shortcomings of the initial reports, other investigators tested for the presence of EBV in HCC obtained from Germany and the UK, utilizing LCM followed by PCR. In this study, the authors reported that tumor samples that were positive for EBV by PCR alone were all negative for EBV when microdissected specimens were used (Junying et al. 2003), demonstrating that EBV is not a potential etiologic agent for HCC.

As with EBV and HCC, the presence of HIV in different brain cell types has been controversial. To resolve this issue, Trillo-Pazos et al. (2003) used LCM to microdissect different types of brain cells, and the tissue lysates were subsequently subjected to PCR. The authors reported that HIV was detected in cells of frontal cortex and basal ganglia; these cells include neurons, and astrocytes, in addition to the previously known macrophages and microglial cells. Collectively, these reports suggest that LCM may be a clinically useful tool to help detect and monitor therapies for virally associated diseases.

2.2.5 Assessment of Gene Copy Numbers

Many laboratories have demonstrated that the presence of consistent genomic alterations in cancer specimens provides important clinical information concerning diagnosis and prognosis. Typically, immunohistochemistry is used to determine expression levels of various genes; however, identification of structural genomic alterations such as gene amplification, deletion, or translocation also has been shown to provide useful clinical information. For example, in breast carcinoma, the amplification and/or overexpression of c-erb-B-2 correlates with poor prognosis (Eissa et al. 1997; Andrulis et al. 1998; Tsuda et al. 1998). In certain cases, gene amplification is a better prognostic indicator than overexpression of the proteins (Ross et al. 1997; Andrulis et al. 1998). This discrepancy between gene copy number and gene expression status has been described for several additional oncogenes and tumor suppressor genes, such as c-myc (Sauter et al. 1995), RB1 (Reissman et al. 1993), and p53 (Cunningham et al. 1992; Eccles et al. 1992; Thompson et al. 1992). In these cases, gene copy number analysis rather than immunohistochemistry appears more valuable.

To date, fluorescence in situ hybridization (FISH) is the best way to identify major chromosomal structural changes in the nuclei. In order to obtain accurate results of FISH, fresh or frozen tissues are required. FISH is difficult in paraffin-embedded tissues because of artifacts from autofluorescence, incomplete nuclei, high background, excessive probe requirement, and poor probe penetration (Qian et al. 1996; McKay et al. 1997). To overcome these limitations of FISH in paraffin-embedded tissues, various protocols to isolate nuclei have been described (Hedley et al. 1983; Heiden et al. 1991). Notwithstanding these technical refinements, an efficient FISH requires pre-treatment in nuclei extracted from formalin-fixed, paraffin-embedded tissues

(Hyytinen et al. 1994). Above all, the use of whole sections to isolate nuclei necessitates that the specimens be a pure collection of cells prior to deriving any meaningful clinical conclusions. FISH using tyramide-based amplification on paraffin specimens has been reported to ease some of these concerns and limitations (De Haas et al. 1996; McKay et al. 1997). Though this technique is useful, simple, and commercial kits are available, the limitations due to autofluorescence, inefficient hybridization, and nonspecific background remain. Moreover, an inaccurate assessment of gene copy numbers can be made because the amplified signals can be large and diffuse.

The above-mentioned limitations of FISH in paraffin-embedded tissues can be overcome largely by using LCM followed by FISH. In a report by DiFrancesco et al. (2000), the authors isolated nuclei from breast cancer tissues that had been stored for more than 1 year following formalin-fixation and paraffin embedding and compared the results with nuclei obtained from concurrently frozen specimens of breast cancer. This study demonstrated that the background is usually cleaner than the touch preparations of viable cells because only the nuclei are present on the slides and cellular RNA is partially lost during the fixation and processing of the specimens. In addition, it was shown that the ability to detect both human cyclin D1 and RB1 simultaneously is equal between touch preparations and LCM specimens and that LCM specimens of breast carcinoma yield nuclei that are very amenable to multicolor FISH analysis. Direct labeling of DNA with fluorescence in multicolor FISH allows one to use many individual probes, limited only by the filter's ability to distinguish similar signals along the entire visible and near-infrared spectrum.

Similarly, LCM has been used to determine gene copy number of proto-oncogene MYCN in neuroblastoma (De Preter et al. 2003). In this study, the authors used quantitative PCR in microdissected samples obtained with LCM and showed that MYCN status correlated well with previous FISH and Southern blot data. Ranging from disseminated disease to spontaneous regression, neuroblastoma (NB) is highly variable in the clinical course. Multiple studies have demonstrated that relevant prognostic indicators include the amplification of the proto-oncogene MYCN (Maris and Matthay 1999). In addition, it has been demonstrated that the status of MYCN amplification predicts the response of infants with stage IV neuroblastoma to combination treatment including chemotherapy, surgery, and local radiation (Schmidt et al. 2000). In fact, the investigators classified patients with MYCN amplification into high risk subgroup within stage IV and administered chemotherapeutic agents in a more dose-intensive induction.

2.2.6 Analysis of Cell Cycle Parameters

Cell cycle analysis of tumor specimens has been shown to provide prognostic information for a number of malignancies. For example, additional prognostic information in subsets of breast cancer cases can be obtained with S-phase fraction determination (Baldetorp et al. 1998; Wong et al. 1999). The accurate estimation of such cell cycle parameters requires tumor cell purity, which is usually difficult to achieve. To address this issue, algorithms have been designed (Baldetorp et al. 1998). Staining with cytokeratin and dual-label gating has also been used to confine the analysis to epithelial cells, and thus, decrease sources of potential errors (McCormick et al. 1998); nevertheless, normal epithelial cells still contaminate the assays.

It has been reported that LCM-dissected specimens from 20-μm sections stained with propidium iodide and subjected to flow cytometric analysis was able to accurately measure ploidy and S-phase fraction (DiFrancesco et al. 2000). The coefficient of variation for the peaks is less than 5%, showing a high technical quality of the specimens. The authors concluded that flow cytometric analysis of LCM specimens for cell cycle parameters is feasible and valuable.

2.2.7 Slide Contamination

Another reported clinical use of LCM includes identification of “floaters” in paraffin-embedded sections. Paraffin tissue block floaters are artifacts that are carried over during the processing of tissues. These small fragments of cross-contaminating tissues can provide a difficult challenge in anatomic pathology. Although clinical and pathologic features allow a satisfactory resolution in many cases, in some cases the precise origin of the suspected tissue floater cannot be accurately established. The clarification is most crucial in cases in which malignant cells are present in the tissue floater and the rest of the tissue is benign. Classically, pathologists faced with potential contaminating tissue floaters will perform the labor-intensive process of examining all other specimens processed in the similar time frame. More recently, DNA finger printing has been used to successfully identify genetic differences in tissue samples. However, this molecular analysis requires a large amount of sample. To circumvent this obstacle, LCM was used to microdissect the floaters that were subsequently analyzed for DNA fingerprints from the extracted DNA. Using this approach, 21 histologically and clinically unresolved cases of tissue floaters were successfully resolved (Hunt et al. 2003).

2.3 Future Clinical Uses: Analysis Of Cellular Proteins

A future potential use of LCM is to procure pure populations of cells for protein analysis. In this regard, several authors (Banks et al. 1999; Ornstein et al. 2000 a, b) demonstrated the feasibility of using LCM to obtain samples in which protein analysis can be performed. For example, renal tissue samples have been microdissected and mass spectrometric sequencing profiles for a number of selected proteins were generated. This study as well as others reported little change in proteins after the application of LCM to the tissue. In fact, studies have shown that proteins recovered from LCM-procured cells retained the ability to bind alpha-1-antichymotrypsin (Ornstein et al. 2000 a, b). Other groups have performed 2-D PAGE analysis of LCM-procured prostatic and esophageal cells and have found that annexin I is commonly underexpressed in prostate and squamous cell carcinoma of the esophagus (Emmert-Buck et al. 2000; Ahram et al. 2003). Curran et al. (2000) reported that 2-D gel electrophoresis followed by mass spectrometry on colonic cancer samples isolated by LCM demonstrated selective loss of nonepithelial proteins when compared to whole tissue samples, thus helping to define a colon cancer-specific protein profile.

A highly innovative use of LCM has been suggested by Paweletz et al. (2001 b). In this review, the authors envision using tissues obtained by LCM to establish tumor-specific protein profile patterns that can be inspected quickly and that correlate with a prognosis or diagnosis. To establish the characteristic protein profiles of varying pathology, surface-enhanced laser desorption ionization (SELDI) spectrometry was used; SELDI is a high-throughput technology for proteomics that is highly sensitive and specific. This technology is easily adaptable to the clinical laboratory and could be used to provide diagnostic as well as prognostic information from minute tissue samples. Another potential application of proteomics to clinical pathology is the construction of protein lysates from LCM-procured cells. This technology facilitates high throughout analysis of protein expression and activation status (Paweletz et al. 2001 a).

2.4 Conclusions

By allowing researchers to isolate and perform molecular analysis on pure populations of cells, tissue microdissection is, in part, revolutionizing translational research. These studies have already yielded important biologic information that is relevant to human disease. In the future, it is likely that microdissection will have a significant and important role in patient care. It is essential that translational scientists work collectively with pathologists and clinicians to realize the full potential of this emerging technology and to improve clinical practice.

References

Ahram M, Flaig MJ, Gillespie JW, Duray PH, Linehan WM, Ornstein DK, Niu S, Zhao Y (2003) Petricoin EF 3rd, and Emmert-Buck MR. Evaluation of ethanol-fixed, paraffin-embedded tissues for proteomic applications. Proteomics 3:413–421

Andrulis IL, Bull SB, Blackstein ME, Sutherland D, Mak C, Sidlofsky S, Pritzker KP, Hartwick RW, Hanna W, Lickley L, Wilkinson R, Qizilbash A, Ambus U, Lipa M, Weizel H, Katz A, Baida M, Mariz S, Stoik G, Dacamara P, Strongitharm D, Geddie W, McCready D (1998) Neu/erb-2 amplification identifies a poor-prognosis group of women with node-negative breast cancer. Toronto Breast Cancer Study Group. J Clin Oncol 16:1340–1349

Baldetorp B, Stal O, Ahrens O, Cornelisse C, Corver W, Falkmer U, Ferno M (1998) Different calculation methods for flow cytometric S-phase fraction: prognostic implication in breast cancer? The Swedish Society of Cancer Study Group. Cytometry 33:385–393

Banks RE, Dunn MJ, Forbes MA, Stanley A, Pappin D, Naven T, Gough M, Harnden P, and Selby PJ (1999) The potential use of laser capture microdissection to selectively obtain distinct population of cells for proteomic analysis-preliminary findings. Electrophoresis 20:689–700

Cooperberg MR, Lubeck DP, Penson DF, Mehta SS, Carroll PR, Kane CJ (2003) Sociodemographic and clinical risk characteristics of patients with prostate cancer within the veterans affairs health care system: data from CaPSURE. J Urol 170:905–908

Cunningham J, Lust JA, Schaid DJ, Bren GD, Carpenter HA, Rizza E, Kovach JS, Thibodeau SN (1992) Expression of p53 and 17p allelic loss in colorectal carcinoma. Cancer Res 52(7):1974–1980

Curran S, McKay JA, McLeod HL, Murray GI (2000) Laser capture microscopy. Mol Pathol 53:64–68

De Haas RR, Verwoerd NP, van der Corput MP, van Gijlswijk RP, Siitari H, Tanke HJ (1996) The use of peroxide-mediated deposition of biotin-tyramide in combination wit h time-resolved fluorescence imaging of europium chelate label in immunohistochemistry and in situ hybridization. J Histochem Cytochem 44:237–243

De Preter K, Vandesompele J, Heimann P, Kockx MM, Van Gele M, Hoebeeck J, De Smet E, Demarche M, Laureys G, Van Roy N, De Paepe A, Speleman F (2003) Application of laser capture microdissection in genetic analysis of neuroblastoma and neuroblastoma precursor cells. Cancer Lett 197:53–61

DiFrancesco LM, Murthy SK, Luider J, Demetrick DJ (2000) Laser capture microdissection-guided fluorescence in situ hybridization and flow cytometric cell cycle analysis of purified nuclei from paraffin sections. Meth Pathol 13:705–711

Eccles DM, Brett L, Lessells A, Gruber L, Lane D, Steel CM, Leonard RC (1992) Overexpression of the p53 protein and allele loss at 17p13 in ovarian carcinoma. Br J Cancer 65:40–44

Eissa S, Khalifa A, el-Gharib A, Salah N, Mohamed MK (1997) Multivariate analysis of DNA ploidy, p53, c-erb-2 proteins, EGFR, and steroid hormone receptors for short-term prognosis in breast cancer. Anticancer Res 17:3091–3097

Emmert-Buck MR, Bonner RF, Smith PD, Chuaqui RF, Zhuang Z, Goldstein SR, Weiss RA, Liotta LA (1996) Laser capture microdissection. Science 274:998–1001

Emmert-Buck MR, Gillespie JW, Paweletz CP, Ornstein DK, Basrur V, Appella E, Wang QH, Huang J, Hu N, Taylor P, Petricoin EF 3rd (2000) An approach to proteomic analysis of human tumors. Mol Carcinog 27:158–165

Hedley DW, Friedlander ML, Taylor IW, Rugg CA, Musgrove CA (1983) Method for analysis of cellular DNA content of paraffin-embedded pathological material using flow cytometry. Cytochemistry 31:1333–1335

Heiden T, Wang N, Bernhard T (1991) An improved Hedley method for preparation of paraffin-embedded tissues for flow-cytometric analysis of ploidy and S phase. Cytometry 12:614–621

Hunt JL, Swalsky P, Sasatomi E, Niehouse L, Bakker A, Finkelstein SD (2003) A microdissection and molecular genotyping assay to confirm the identity of tissue floaters in paraffin-embedded tissue blocks. Arch Pathol Lab Med 127:213–217

Hyytinen E, Visakorpi T, Kallioniemi OP, Isola JJ (1994) Improved technique for analysis of formalin-fixed, paraffin embedded tumors by fluorescence in situ hybridization. Cytometry 16:93–99

Junying J, Herrmann K, Davies G, Lissauer D, Bell A, Timms J, Reynolds GM, Hubscher SG, Young LS, Niedobitek G, Murray PG (2003) Absence of Epstein-Barr virus DNA in the tumor cells of European hepatocellular carcinoma. Virology 306:236–243

Maris JM, Matthay KK (1999) Molecular biology of neuroblastoma. J Clin Oncol 17:2264–2279

McCormick SR, Peters AA, Schrauth JB (1998) Flow cytometric DNA analysis with cytokeratin gating of formalin-fixed deparffinized breast cancer nuclei. Am J Clin Pathol 110:227–237

McKay JA, Murray GI, Keith WN, McLeod HL (1997) Amplification of fluorescent in situ hybridization signals in formalin fixed paraffin wax embedded section of colon tumor using biotinylated tyramide. Mol Pathol 50:322–325

Miura K, Bowman ED, Simon R, Peng AC, Robles AI, Jone RT, Katagiri T, He P, Mizukami H, Charboneau L, Kikuchi T, Liotta LA, Nakamura Y, Harris CC (2002) Laser capture microdissection and microarray expression analysis of lung adenocarcinoma reveals tobacco smoking- and prognosis-related molecular profiles. Cancer Res 62:3244–3250

Orba Y, Tanaka S, Nishihara H, Kawamura N, Itoh T, Shimizu M, Sawa H, and Nagashima K (2003) Application of laser capture microdissection to cytologic specimens for the detection of immunoglobulin heavy chain gene rearrangement in patients with malignant lymphoma. Cancer 99:198–204

Ornstein DK, Englert C, Gillespie JW, Paweletz CP, Linehan WM, Emmert-Buck MR, and Petricoin EF 3rd (2000 a) Characterization of intracellular prostate-specific antigen from laser capture microdissected benign and malignant prostatic epithelium. Clin Cancer Res 6:353–356

Ornstein DK, Gillespie JW, Paweletz CP, Duray PH, Herring J, Vocke CD, Topalian SL, Bostwick DG, Linehan WM, Petricoin EF 3rd, Emmert-Buck MR (2000 b) Proteomic analysis of laser capture microdissected human prostrate cancer and in vitro prostate cell lines. Electrophoresis 21:2235–2242

Paradis V, Dargere D, Bonvoust F, Rubbia-Brandt L, Ba N, Bioulac-Sage P, Bedossa P (2000) Clonal analysis of micronodules in virus C-induced liver cirrhosis using laser capture microdissection (LCM) and HUMARA assay. Lab Invest 80(10):1553–1559

Paweletz CP, Charboneau L, Bichsel VE, Simone NL, Chen T, Gillespie JW, Emmert-Buck MR, Roth MJ, Petricoin III EF, Liotta LA (2001 a) Reverse phase protein microarrays which capture disease progression show activation of pro-survival pathways at the cancer invasion front. Oncogene 20:1981–1989

Paweletz CP, Liotta LA, Petricoin III EF (2001 b) New technologies for biomarker analysis of prostate cancer progression: laser capture microdissection and tissue proteomics. Urology 57:160–163

Qian J, Bostwick DG, Takahashi S, Borell TJ, Brown JA, Lieber MM, Jenkins RB (1996) Comparison of fluorescence in situ hybridization analysis of isolated nuclei and routine histological sections from paraffin-embedded prostatic adenocarcinoma specimens. Am J Pathol 149:1193–1199

Reissmann PT, Koga H, Takahashi R, Figlin RA, Holmes EC, Piantadosi S, Cordon-Cardo C, Slamon DJ (1993) Inactivation of the retinoblastoma susceptibility gene in non-small-cell lung cancer. The Lung Cancer Study Group. Oncogene 8:1913–1919

Ross JS, Sheehan C, Hayner-Buchan AM, Ambros RA, Kallakury BV, Kaufman R, Fisher HA, Muraca PJ (1997) HER-2/neu gene amplification status in prostate cancer by fluorescence in situ hybridization. Human Pathol 28:827–833

Sauter G, Carroll P, Moch H, Kallioniemi A, Kerschmann R, Narayan P, Mihatsch MJ, Waldman FM (1995) C-myc copy number gains in bladder detected by fluorescence in situ hybridization. Am J Pathol 146:1131–1139

Schmidt ML, Lukens JN, Seeger RC, Brodeur GM, Shimada H, Gerbing RB, Stram DO, Perez C, Haase GM, Matthay KK (2000) Biologic factors determine prognosis in infants with stage IV neuroblastoma: a prospective children's cancer group study. J Clin Oncol (18(6):1260–1268

Sugawara Y, Mizugaki Y, Uchida T, Torii T, Imai S, Makuuchi M, Takada K (1999) Detection of Epstein-Barr virus (EBV) in hepatocellular carcinoma tissue: a novel EBV latency characterized by the absence of EBV-encoded small RNA expression. Virology 256:196–202

Sugawara Y, Makuuchi M, Takada K (2000) Detection of Epstein-Barr virus DNA in hepatocellular carcinoma tisseuse from hepatitis C-positive patients. Scand J Gastroentrerol 35:981–984

Thompson AM, Anderson TJ, Condie A, Prosser J, Chetty U, Carter DC, Evans HJ, Steel CM (1992) p53 allele losses, mutations, and expression in breast cancer and their relationship to clinico-pathological parameters. Int J Cancer 50:528–532

Trillo-Pazos G, Diamanturos A, Rislove L, Menza T, Chao W, Belem P, Sadiq S, Morgello S, Sharer L, Volsky DJ (2003) Detection of HVI-1 DNA in microglia/macrophages, astrocytes, and neurons isolated from brain tissues with HIV-1 encephalitis by laser capture microdissection. Brain Pathol 13:144–154

Tsuda H, Sakamaki C, Tsugane S, Fukutomi T, Hirohashi S (1998) A prospective study of the significance of gene and chromosome alterations as prognostic indicators of breast cancer patients with lymph node metastases. Breast Cancer Res Treat 48:21–32

Wong SW, Rangan AM, Bilous AM, Boyages J, Gebski V, Benson EM (1999) The value of S-phase and DNA ploidy analysis as prognostic markers for node-negative breast cancer in the Australian setting. Pathology 31:90–94

3 Tissue Processing

Gallya Gannot and John W. Gillespie

3.1 Introduction

With the completion of the human genome sequence and the development of high-throughput technologies for gene and protein expression analysis, there has been a tremendous increase in the rate of data acquisition (Schena et al. 1995; DeRisi et al. 1996; Anderson et al. 1998; Lipshutz et al. 1999; Emmert-Buck et al. 2000) making a comprehensive study of disease processes at the molecular level feasible.

Several experimental systems for molecular profiling are available, including animal models, human cells in vitro (Zhang et al. 1997; Celis et al. 1999), and ex vivo specimens. While cells in vitro will always be very useful, their molecular profiles may not always be reflective of the molecular profiling seen in vivo (Ornstein et al. 2000), since there may be changes in gene and protein expression induced by the culture conditions. Since ex vivo samples analyzed directly from patients most closely represent conditions in vivo, it is likely that molecular profiles derived from them will give pertinent information on the disease processes, including pathogenesis and also potential clinically important information such as diagnostic, prognostic markers and therapeutic targets. The ability to analyze gene and protein expression from tissue samples will be critically dependent on how they are processed. In the future, it will be important that tissue fixation and embedding methods integrate both excellent histology and optimal molecular preservation. In this way, the information obtained not only from the histology, but also from molecular analysis will provide insight into molecular pathogenesis of disease processes. This chapter will discuss various aspects of handling clinical tissue samples emphasizing not only maintenance of adequate histology, but also optimization of preservation of DNA, RNA, and proteins for molecular analyses.

Principles and Practice
Emmert-Buck et al. (Eds.):
Dissecting the Molecular Anatomy of Tissue

3.2 Clinical Aspects

Pathologic examination of tissue sections involves patient care; therefore, it is essential that an accurate diagnosis be rendered. Traditionally, biopsy specimens and tissue from surgical resection are fixed in 10% normal buffered formalin and embedded in paraffin. Tissue sections are then cut onto glass slides, which are hematoxylin and eosin-stained. Since the histopathologic criteria for diagnosis of a variety of diseases in tissue is based on formalin fixation, it has become the gold standard for diagnostic histopathology.

Tissue is typically frozen either for an intraoperative consultation, or for molecular analysis. For intraoperative consultation, tissue sections are cut onto glass slides, which are then hematoxylin and eosin-stained. For frozen section diagnosis, the histology is usually adequate to identify the presence of a particular lesion; however, subtle histologic details may not be apparent. Therefore, the final diagnosis is typically rendered on other portions of the specimen that have been formalin-fixed. At present, molecular analysis is most commonly performed on "extra tissue" from the lesion, which is not required for diagnosis and staging. This is typically snap-frozen at the time of surgery and then used for a variety of molecular analyses.

3.3 Sample Acquisition and Initial Tissue Handling

If molecular profiling is a consideration for a specimen, then it is essential that careful planning and close coordination between clinicians, pathologists, and researchers take place prior to the procedure. After removal of tissue from the body, initial handling entails transportation from the operating room to the pathology department, inking surgical margins, and cutting the tissue into sufficiently small sections. During this period, degradation of biomolecules can occur, particularly proteins and RNA. Thus, there are essentially three major considerations of initial tissue handling to minimize degradation of DNA, RNA, and proteins including temperature, time of handling, and specimen size. To decrease the rate of biomolecular degradation, the tissue should be placed immediately on ice. All the steps of tissue preparation prior to fixation or snap-freezing, including surgical margin inking and sectioning, should proceed as quickly as possible. Finally, the smaller the tissue pieces are grossly sectioned, the quicker the tissue will be fixed or frozen.

3.4 Tissue Fixation and Snap-Freezing

Historically, with the establishment of the microscope as a diagnostic tool in the early 1800s, information of diagnostic value could be obtained from the examination of diseased tissue (Leong et al. 1996). Early on, improvements of the microscope were the main objective of microscopists. By the end of the nineteenth century, the importance of quality of specimens that were being observed was realized. Attention was then focused on developing fixatives that would preserve cell and tissue constituents in the state that they appear in life while allowing preparative procedures. Amongst these fixatives, formaldehyde was first applied to tissue samples by Ferdinand Blum (Fox et al. 1985). It was observed to cause less tissue artifact than alcohols that were used prior to that time. Ten percent formalin has been the most commonly used fixative in pathology departments today. A list of various fixatives is shown in Table 3.1. In 1880, wax embedding and the advent of fixation and tissue processing gave rise to an increased interest in examination of biopsies.

Snap-freezing, or fixation and embedding are the two different ways in which tissue is typically processed. Figure 3.1 shows the steps of tissue handling for both freezing, as well as fixation and embedding (Leiva et al. 2003). Traditionally, tissue which is to be used for molecular analysis is frozen, whereas tissue which is to be examined histologically is fixed and embedded. Since the diagnosis of many diseases is usually performed based on histologic examination, fixation and embedding methods for clinical tissue specimens in pathology departments do not generally prioritize preservation of DNA, RNA, and proteins. The focus has been to achieve optimal histology. Therefore, standard surgical pathology procedures are inadequate for use with high-throughput molecular technologies (Ben-Ezra et al. 1991; Klimecki et al. 1994). Traditionally, optimal molecular preservation in tissue has been achieved through snap-freezing. The disadvantages of this method are poor histology (Fend et al. 1999; Gillespie et al. 2000) as well as inconvenient storage and handling. In the future, it is important that tissue fixation and embedding methods should integrate both excellent histology and optimal molecular preservation. In this way, the information obtained not only from the histology, but also from molecular analysis will provide insight into etiology of disease processes. There are many steps in fixation and tissue processing in which different physical and chemical conditions can be varied to optimize both histology and molecular preservation (Fig. 3.1). Ultimately, it is critical that a "universal protocol" be developed for fixation and embedding which would not only retain excellent histology and allow convenient storage and handling, but would also ideally preserve biomolecules.

Table 3.1. Fixatives that would preserve cell and tissue constituents (Pearse 1968; Bancroft and Stevens 1990; Kiernan 1990; Prophet et al. 1992; Srinivasan et al. 2002)

Group	Name of fixative	Special features
Liquids	Formaldehyde	1. Most commonly used fixative for tissue specimens for pathology 2. Cross-linking agent 3. Excellent retention of tissue morphology 4. Very limited for molecular analysis
	Glutaraldehyde	5. Penetrates tissue slowly 6. Fixation occurs by cross-linking of proteins that is more extensive than formaldehyde 7. Slight extraction of nucleic acids 8. No carbohydrate fixation 9. No lipid fixation 10. Preferable fixative for electron microscopy 11. Fixation is performed in cold temperature
	Acrolein	1. A brief exposure is necessary (15 min) 2. Cross-links proteins 3. Very unstable and is not used in routine fixation procedures 4. Used for electron microscopy 5. Very toxic
	Genipin	1. Naturally occurring cross-linking agent 2. Increases tissue resistance to collagenase degradation 3. May form intra-and intermolecular cross-links within collagen
	Carnoy's (ethanol 60%, chloroform 30%, glacial acetic acid 10%)	1. Studies suggest that it gives optimal preservation of nucleic acids
	Methacarn (methanol 60%, chloroform 30%, glacial acetic acid 10%)	1. Excellent preservative of RNA 2. Superior to formalin in retaining antigen immunoreactivity in tissue and does not require antigen retrieval
	Absolute alcohol	1. Precipitating agent 2. Preserves glycogen 3. May cause distortion of nuclear detail 4. Shrinks cytoplasm
	Cold acetone	1. Precipitating agent 2. Preserves tissue morphology 3. Maintains immunoreactivity for labile proteins 4. Preserves DNA and RNA for analysis 5. Does not preserve glycogen 6. Shrinks cytoplasm

Table 3.1 (continued)

Group	Name of fixative	Special features
	HOPE fixative (Hepes-glutamic acid buffer-mediated organic solvent protective effect)	1. Compatible for complete pathologic analysis including retention of features for immuno-histochemistry and molecular pathology
	Acetic acid	1. Rapid tissue penetration 2. No protein fixation 3. Precipitates nucleic acids 4. No carbohydrate fixation 5. No lipid fixation 6. Used to fix only small blocks. Is not used alone as causes extensive shrinkage of tissue
	Hydroxyadipaldehyde and crotonaldehyde	1. Used in electron microscopy 2. Are inferior to glutaraldehyde in preserving tissue structure
	Trichloracetic acid	1. Rapid penetration of tissue 2. Coagulates proteins 3. Extracts nucleic acids 4. No carbohydrate fixation 5. No lipid fixation 6. Not used as a single fixative. Sometimes used as a decalcifying solution
	Potassium permanganate	Useful for demonstrating membrane structures
Solids	Mercuric chloride	1. Coagulates proteins 2. Coagulates nucleic acids 3. No carbohydrate fixation 4. Reacts with lipids 5. Used in combination with other fixatives 6. Shrinks tissue
	Chromium compounds	1. Depends on pH. When pH is less than 3.5: 1.1 Is rarely used 1.2 Slow tissue penetration 1.3 Coagulation of proteins 1.4 Coagulation and hydrolysis of nucleic acids 1.5 Oxidation of carbohydrates 1.6 Oxidation of lipids 2. When pH is higher than 3.5: 2.1 Is rarely used 2.2 Rapid tissue penetration 2.3 No coagulation of proteins 2.4 Slight extraction of nucleic acids 2.5 No effect on carbohydrates

Table 3.1 (continued)

Group	Name of fixative	Special features
	Osmium tetroxide	1. Slow tissue penetration 2. Slight extraction of proteins and some cross-linking 3. Causes breakdown of proteins 4. Slight extraction of nucleic acids (reacts with pyrimidine) 5. Oxidation of lipids 6. Used for specific applications and in electron microscopy 7. Requires small samples
	Picric acid	1. Slow tissue penetration 2. Precipitates proteins (forms crystalline picrates) 3. Partly hydrolyses nucleic acids 4. No carbohydrate fixation 5. No lipid fixation 6. Usually used with other fixatives. Tissue needs to be thoroughly washed due to continuous picric acid reaction with tissue 7. Causes shrinkage of tissue 8. Used for electron microscopy and immunocytochemistry
Temperature	Heat	1. Used for antigen retrieval for immunohistochemistry 2. Used to facilitate penetration of fixatives 3. May produce good morphological results when the tissue is heated in a constant volume bath to obtain accurate temperature 4. Will denature proteins and melt lipids
	Freezing	1. Best method for preservation proteins and nucleic acids 2. Poor preservation of tissue morphology 3. Difficult procedure for routine fixation since the freezing needs to be quick and the tissue small (around 2 mm thick)

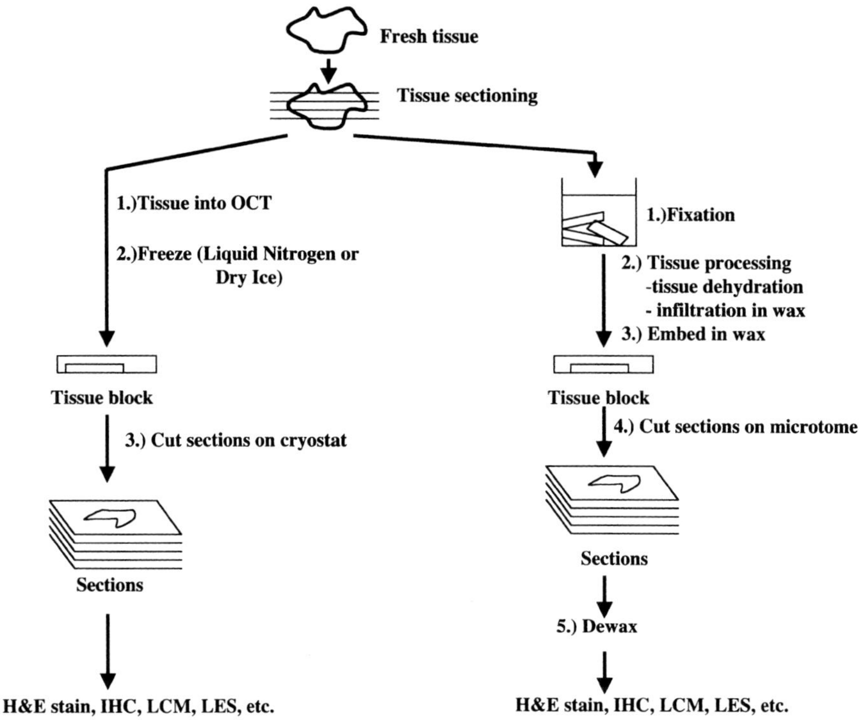

Fig. 3.1. Steps of tissue handling for frozen and fixation and embedding. There are many more steps for tissue fixation and embedding, which can be impacted for preservation of DNA, RNA, and proteins than with freezing. Before undertaking a study, an investigator will need to determine the choice of method based on the type of analysis to be performed and the tissue type

3.5 Tissue Embedding

The purpose of embedding is to allow convenient storage of frozen or fixed tissue in a solid medium, to prevent the tissue from drying out, and allow the tissue to be cut uniformly into very thin sections. For frozen samples, a piece of fresh tissue is placed in a cryomold which is then filled with optimal cutting temperature (OCT) embedding media, then frozen on dry ice or liquid nitrogen to form a block. The block is then stored at –80 or –120 °C. For fixed tissue, after fixation the specimen is typically processed through multiple steps of dehydration in increasing concentrations of ethanol, and a final step in xylene. This is followed by infiltration of the tissue with an embedding medium, usually a wax. After infiltration, the tissue is then placed in a mold and is embedded with the same medium to form a block. The block can then be stored at room temperature. The characteristics of an embedding medium include the requirement that it be inert, be able to repel moisture, be able to readily penetrate tissue, cut a tissue section of uniform thickness, and maintain its integrity at room temperature. Various types of waxes fulfill these conditions including the routinely used embedding medium, paraffin wax. Tissue sections are then cut onto glass slides 5–10 μm thick in a cryostat or on a microtome from frozen or wax-embedded blocks, respectively.

3.6 Evaluation of New Methods for Handling Clinical Tissue Specimens: Prostate Cancer as an Example of Tissue Handling for Molecular Analysis

3.6.1 New Methods of Fixation/Embedding

Diverse methods of tissue fixation and embedding have been evaluated not only to produce excellent histology, but also to promote the elucidation of the gene and protein expression profiles. Several methods utilizing alcohol-based fixation have shown promising results (Goldsworthy et al. 1999; Gillespie et al. 2000; Shibutani et al. 2000). It is our belief that any new fixation protocol that is to be evaluated for clinical specimens should be compared to 10% formalin fixation for histologic characteristics and with snap-frozen tissue for recovery of DNA, RNA, and proteins.

The development of an optimized protocol to fix and embed tissue that will not only preserve biomolecules, but also retain optimal histology as seen in formalin fixation and paraffin embedding is an active area of investigation in our laboratory. Early work in our lab compared the recovery of RNA in murine tissue for a variety of alcohol-based and cross-linking fixatives. It was determined that the alcohol-based fixatives showed better recovery of RNA than the cross-linking fixatives (data unpubl.). Subsequent work involving pathologists from Johns Hopkins University and National Institutes of Health in a blinded study compared the histology of small pieces of prostate tissue derived from prostatectomy specimens for the same fixatives (Gillespie et al. 2000). Besides giving histology comparable to formalin, 70% ethanol was selected for future studies over other fixatives since it gave very good recovery of RNA and is relatively inexpensive and nontoxic. The work with prostate described below represents the approach we have used to evaluate a tissue processing method. By no means is this the final method. We feel that it will be an evolving process with many groups looking at other approaches and other organ systems besides prostate until the best method is achieved. Any approach that one uses should compare histology with standard formalin fixation and experiments comparing DNA, RNA, and protein recovery with snap-frozen tissue.

3.6.2 Nonformalin Tissue Fixation

Since the work described above, we have processed over 100 prostatectomy specimens as whole-mounts. In the operating room, immediately after the prostates are resected, they are placed immediately on ice. The prostates are then transported to the pathology department where the surgical margins are inked. Each prostate is then transverse-sectioned into tissue sections, which are 3–5 mm thick and are oriented from the apex to the base of the gland (Fig. 3.2, see also color illustration at the end of the book). The sections are then fixed in 70% ethanol at 4 °C and embedded in paraffin to form tissue blocks. The fixation and embedding protocols are described in detail in Sections 11.1.1 and 11.1.2, respectively, of the Protocols section. From each of the blocks, a section is cut which is then hematoxylin and eosin-stained. The entire prostate is then examined by a pathologist with the subsequent generation of a surgical pathology report.

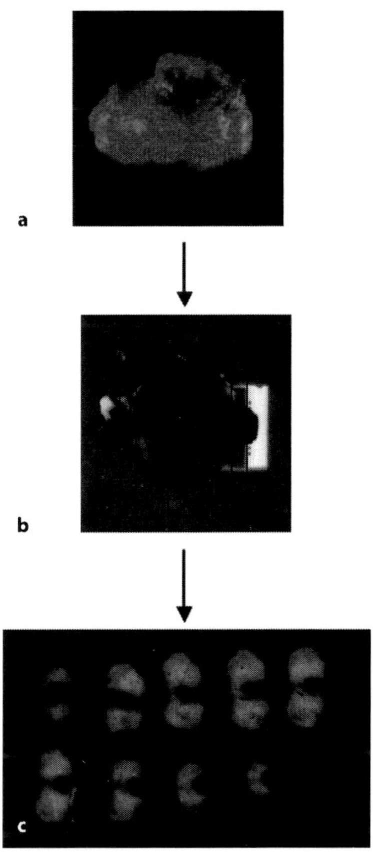

Fig. 3.2. Steps of initial prostate handling. **a** Fresh prostatectomy specimen is recovered immediately after resection. **b** Prostate after being inked with surgical ink or acrylic paint in order to delineate the surgical margin and to maintain the orientation. Lines, spaced every 3–5 mm, show location of transverse sectioning perpendicular to the urethra from base to apex. **c** Sections of the prostate after being transverse sectioned with an extra large scalpel blade. These are then placed into cassettes, which are placed into fixative (see color illustration at the end of the book)

3.6.3 Comparison of Histology and Recovery of DNA, RNA, and Proteins

In a small subset of these cases, in addition to fixation in 70% ethanol and paraffin-embedding, tissue sections were also snap-frozen, ethanol-fixed and polyester wax-embedded, and formalin-fixed and paraffin-embedded in order to do a more thorough comparison of the histopathology and recovery of DNA, RNA, and proteins (Gillespie et al. 2000). The formalin and 70%

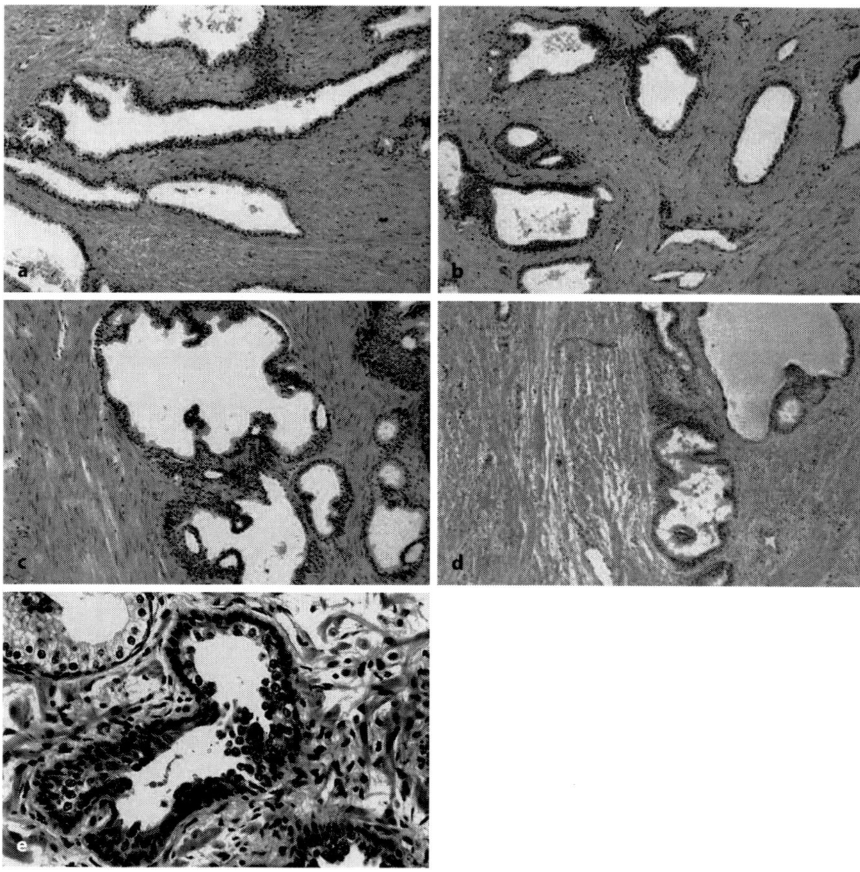

Fig. 3.3. Comparison of the histological quality of 5-μm-thick sections of prostate tissue stained with H & E. **a–d** Normal prostate. Original magnification, ×100. **a** Formalin-fixed, paraffin-embedded. **b** 70% Ethanol-fixed, paraffin-embedded. **c** 70% Ethanol-fixed, polyester wax-embedded. **d** Frozen. Note the similar staining quality and architecture for the formalin-fixed and ethanol-fixed tissues, which are superior to the frozen. **e** High-grade prostatic intraepithelial neoplasia from ethanol-fixed, paraffin-embedded prostate tissue showing nuclear overlap and fine nuclear detail, including nucleolar prominence. H & E stained, original magnification ×200 (see color illustration at the end of the book)

ethanol-fixed samples show comparable histology and both are superior to the frozen sample (Fig. 3.3, see also color illustration at the end of the book). Normal, hyperplastic epithelium, prostatic intraepithelial neoplasia (PIN), and adenocarcinoma are readily distinguished from each other. By PCR analysis and recovery of total DNA, we observed that recovery of DNA is comparable for the frozen and the ethanol-fixed tissues, which are superior to the formalin-fixed tissue (Fig. 3.4). Using total RNA analysis on a denaturing gel, recovery of RNA is best from snap-frozen tissue, followed by ethanol-

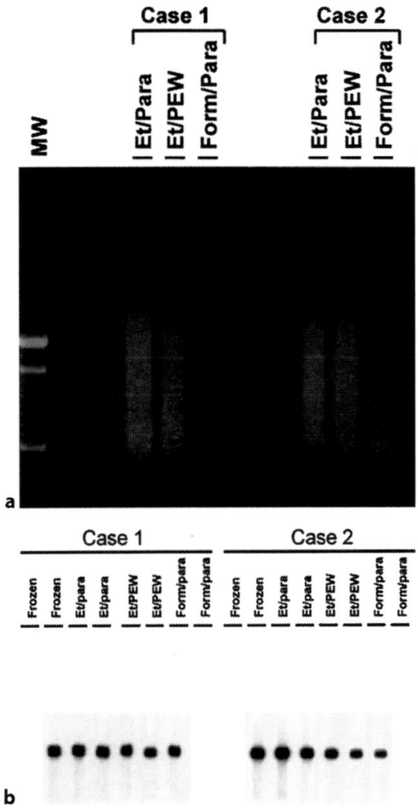

Fig. 3.4. **a** Analysis of total DNA quality in prostate tissue that was either ethanol-fixed and paraffin-embedded (*Et/Para*), ethanol-fixed and polyester wax-embedded (*Et/PEW*), or formalin-fixed and paraffin-embedded (*Form/Para*). Each sample was loaded onto a 1% agarose gel. DNA was visualized with ethidium bromide staining. **b** PCR-based analysis of DNA quality in prostate tissue that was either snap-frozen, ethanol-fixed and paraffin-embedded (*Et/Para*), ethanol-fixed and polyester wax-embedded (*Et/PEW*), or formalin-fixed and paraffin-embedded (*Form/Para*). The PCR product from amplification of microsatellite marker D17S926 was electrophoresed on a 6% denaturing acrylamide gel and visualized by autoradiography. All samples were analyzed in duplicate

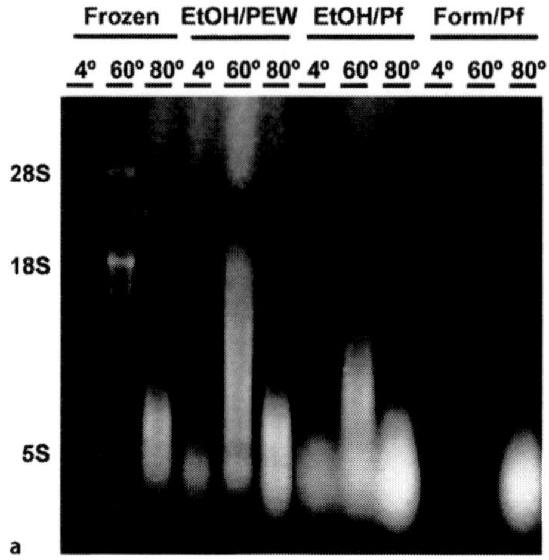

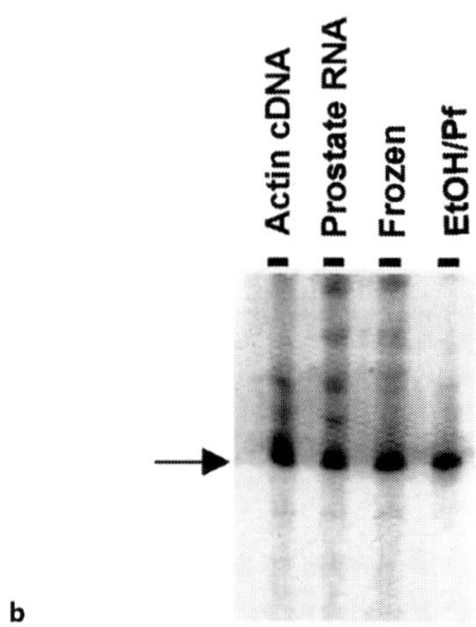

Fig. 3.5. **a** Denaturing agarose gel of total RNA from prostate tissue sections that were either frozen, ethanol-fixed, polyester wax-embedded (*EtOH/PEW*), ethanol-fixed, paraffin-embedded (*EtOH/Pf*), or formalin-fixed, paraffin-embedded (*Form/Pf*). **b** RT-PCR for actin. Approximately 15,000 epithelial cells were microdissected from frozen or ethanol-fixed, paraffin-embedded (*EtOH/Pf*) prostate sections. Actin cDNA and prostate RNA from cell culture are included as positive controls. The *arrow* indicates the 220-bp band corresponding to actin

fixed tissue, and the least RNA is recovered from formalin-fixed tissue (Fig. 3.5 A). For ethanol-fixed tissue, RNA recovery can be improved by pre-incubation of the tissue lysates for 20 min at 60 °C prior to RNA extraction. Furthermore, a comparable signal for snap-frozen and ethanol-fixed tissue can be attained by RT-PCR (Fig. 3.5 B). The RT-PCR protocol is described in detail in Section 11.5.2.1 of the Protocols section. Using 1-D PAGE, recovery of total protein is comparable for the frozen and the ethanol-fixed pre-

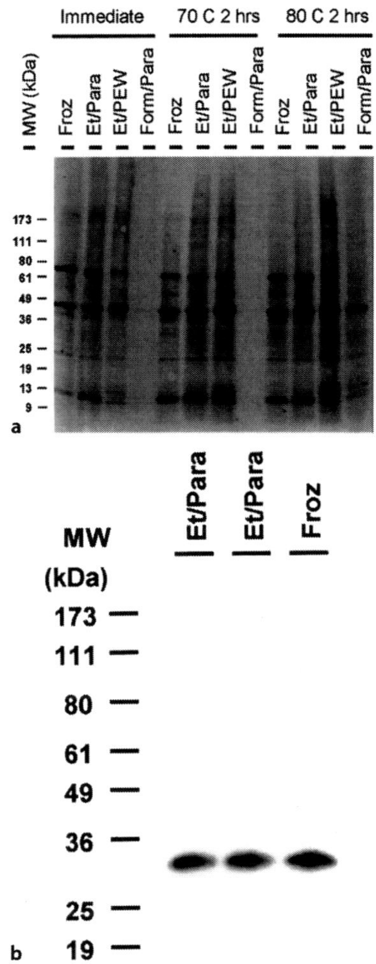

Fig. 3.6. Protein recovery from prostate tissue sections. **a** Samples were either snap frozen (*Froz*), ethanol-fixed and paraffin-embedded (*Et/Para*), ethanol-fixed and polyester wax-embedded (*Et/PEW*), or formalin-fixed and paraffin-embedded (*Form/Para*). **b** Immunoblot for PSA. Approximately 35,000 prostate epithelial cells were microdissected from histological sections using LCM. Two separate microdissections from the ethanol-fixed, paraffin-embedded sample were performed to assess reproducibility

parations which are both superior to formalin fixation (Fig. 3.6 A). Recovery of protein from the formalin-fixed sample is improved by pre-incubation of the tissue lysate at 80 °C prior to protein extraction. Also, ethanol-fixed and frozen tissues gave comparable signals by immunoblot (Fig. 3.6 B). The immunoblot protocol is described in detail in Section 11.5.3.2 of the Protocols section. Ethanol-fixed tissue can be applied to a recently developed technique, layered expression scanning (LES; Englert et al. 2000), in which the entire protein content has been passed through ten membrane layers. After staining with Coomassie blue, all layers also show retention of the anatomic structures seen within the original tissue section (Fig. 3.7). This strategy for evaluating clinical tissue samples has shown that ethanol fixation and paraffin embedding allows for histology similar to formalin-fixation with much improved preservation of biomolecules over formalin-fixation.

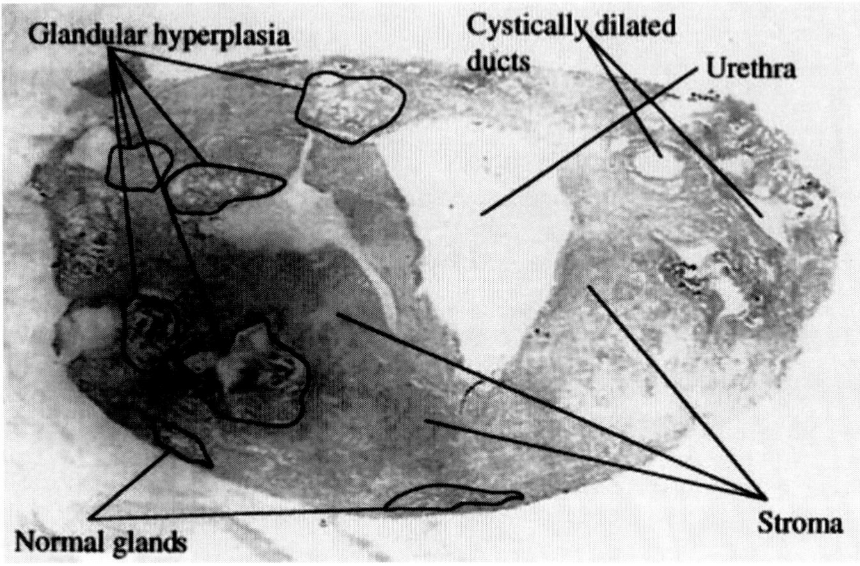

Fig. 3.7. Layered expression scanning using an ethanol-fixed, paraffin-embedded prostate which has been transferred onto nitrocellulose and stained with Coomassie blue. The *labeled regions* demonstrate that the protein staining pattern retains characteristic prostate structures

3.7 Conclusions and Future Directions

With the completion of the Human Genome Project, and the development of high-throughput technologies to obtain gene and protein expression profiles, it has become feasible to study profiles of the molecular changes that may be the basis of disease. However, the amount of molecular information that can be obtained from clinical tissue samples depends on how the samples are handled and processed. We have demonstrated that 70% ethanol fixation of prostate preserves histopathology and also allows for recovery of DNA, RNA, and proteins for subsequent molecular profiling studies. We are continuing to investigate the types of molecular analyses that can be performed on ethanol-fixed samples, including the use of ethanol-fixed tissue for use on cDNA microarrays, as well as the stability of biomolecules and histology from 70% ethanol-fixed tissue over time.

Future efforts will include comparing many different fixatives to improve the histology and preservation of biomolecules and examining the influence of the surgical procedure on molecular profiling. Reversible cross-linking agents in the fixative may preserve biomolecules in tissue for a long period of time by constricting the movement of biomolecules and inhibiting RNase and proteinase activity.

Ultimately, molecular profiling success depends on using a general strategy for evaluating clinical tissue specimens, with new fixation/embedding methods allowing investigators to perform high throughput molecular analysis on selected cell populations of a sample. Continued improvement of tissue processing methodologies will be a critical step toward ultimately determining the molecular anatomy of normal and diseased human cell types.

Acknowledgements. This project has been funded in whole or in part with Federal funds from the National Cancer Institute, National Institutes of Health, under Contract No. N01-C0-12400.

References

Anderson N, Anderson NG (1998) Proteome and proteomics: new technologies, new concepts and new words. Electrophoresis 19:1853–1861

Anderson L, Seilhamer J (1997) A comparison of selected mRNA and protein abundances in human liver. Electrophoresis 18:533–537

Bancroft JD, Stevens A (1990) Theory and practice of histological techniques, 3rd edn. Churchill Livingstone, Edinburgh

Ben-Ezra J, Johnson DA, Rossi J, Cook N, Wu A (1991) Effect of fixation on the amplification of nucleic acids from paraffin-embedded material by the polymerase chain reaction. J Histochem Cytochem 39:351–354

Celis J, Rasmussen HH, Celis P, Bodil P, Lauridsen JB, Ratz G, Hein B, Ostergaard M, Orntoft T, Celis J (1999) Short-term culturing of low-grade superficial bladder transitional cell carcinoma leads to changes in the expression levels of several proteins involved in key cellular activities. Electrophoresis 20:355–361

DeRisi JL, Penland L, Brown ML, Meltzer PS, Ray M, Chen Y, Su YA, Trent M (1996) Use of a cDNA microarray to analysis gene expression pattern in human cancer. Nat Gen 14:457–460

Emmert-Buck MR, Strausberg RL, Krizman DB, Bonaldo MF, Bonner RF, Bonner RF, Bostwick DG, Brown MR, Buetow KH, Chuaqui RF, Cole KA, Duray PH, Englert CR, Gillespie JW, Greenhut S, Grouse L, Hillier LW, Katz KS, Klausner RD, Kuznetzov V, Lash AE, Lennon G, Linehan MW, Liotta LA, Marra MA, Munson PJ, Ornstein DK, Prabhu VV, Prange C, Schuler GD, Bento Soares M, Tolstoshev CM, Vocke CD, Waterston RH (2000) Molecular profiling of clinical tissue specimens: feasibility and applications. Am J Pathol 156:1109–1115

Englert CR, Baibakov GV, Emmert-Buck MR (2000) Layered expression scanning: rapid molecular profiling of tumor samples. Cancer Res 60:1526–1530

Fend F, Quintanilla-Martinez L, Kumar S, Beaty MW, Blum L, Sobara L, Jaffe EF, Raffeld M (1999) Composite low grade B-cell lymphomas with two immunophenotypically distinct cell populations are true biclonal lymphomas: a molecular analysis using laser capture microdissection. Am J Pathol 154:1857–1866

Fox CH, Johnson FB, Whiting J, Roller PP (1985) Formaldehyde fixation. J Histochem Cytochem 8:845–853

Gillespie JW, Best CJM, Bichsel VE, Cole KA, Greenhut SF, Hewitt SM, Ahram M, Gaithright YB, Merino MJ, Strausberg RL, Epstein JI, Hamilton SR, Gannot G, Baibakova GV, Calvert VS, Flaig MJ, Chuaqui RF, Herring JC, Pfeifer J, Petricoin

EF, Linehan WM, Duray PH, Bova GS, Emmert-Buck MR (2000) Evaluation of non-formalin tissue fixation for molecular profiling studies. Am J Pathol 156:1109–1115

Goldsworthy SM, Stockton PS, Trempus CS, Foley JF, Maronpot RR (1999) Effects of fixation on RNA extraction and amplification from laser capture microdissected tissue. Mol Carcinog 25:86–91

Kiernan JA (1990) Histological and histochemical methods, theory and practice, 2nd edn. Pergamon Press, New York

Klimecki WT, Futscher BW, Dalton WS (1994) Effects of ethanol and paraformaldehyde on RNA yield and quality. Biotechniques 16:1021–1023

Leiva IM, Emmert-Buck MR, Gillespie JW (2003) Handling of clinical tissue specimens for molecular profiling studies. Curr Issues Mol Biol 5:27–35

Leong ASY, James CL, Thomas AC (1996) Brief history of surgical pathology. In: Handbook of surgical pathology. Churchill Livingstone, New York, pp 1, 2

Lipshutz R, Fodor SPA, Gingeras TR, Lockhart DJ (1999) High density synthetic oligonucleotide arrays. Nat Genet 21:20–24

Ornstein DK, Gillespie JW, Paweletz CP, Duray PH, Herring J, Vocke CD, Topalian SL, Bostwick DG, Linehan WM, Petricoin EF, Emmert-Buck MR (2000) Proteomic analysis of laser capture microdissected human prostate cancer and in vitro prostate cell lines. Electrophoresis 21:2235–2242

Pearse AGE (1968) Histochemistry, theoretical and applied, 3rd edn. Churchill Livingstone, London

Prophet ED, Mills B, Arrington JB, Sobin LH (1992) Laboratory methods in histotechnology. Armed Forces Institute of Pathology, American Registry of Pathology, Washington, DC

Schena M, Shalon D, Davis RW, Brown PO (1995) Quantitative monitoring of gene expression patterns with a complementary DNA microarray. Science 270:467–469

Shibutani M, Unemaya C, Miyazaki K, Teyoda K, Hirose M (2000) Methacam fixation: a novel tool for analysis of gene expressions in paraffin-embedded tissue specimens. Lab Invest 80:199–208

Srinivasan M, Sedmak D, Jewell S (2002) Effect of fixatives and tissue processing on the content and integrity of nucleic acids. Am J Pathol 161:1961–1971

Zhang L, Zhou W, Velculescu VE, Kern SE, Hruban RH, Hamilton SR, Vogelstein B, Kinzler KW (1997) Gene expression profiles in normal and cancer cells. Science 276:1268–1272

4 Sample Management and Tracking

G. Steven Bova

4.1 Introduction

The key value provided by tissue microdissection is the separation of whole tissues into singularly defined component parts for molecular analysis. Optimal tissue microdissection occurs when the dissector possesses a singular morphological definition of target cells to be dissected, and has enough technical knowledge and ability using the dissecting tool to apply this understanding to obtain target cell material exclusive of nontarget cell material. However, obtaining pure target cells is a necessary, but not sufficient step toward converting the pure target cell material to robustly analyzable biologic data. Equally necessary for analysis are a workflow that incorporates quality control and capture of accurate metadata sufficient for proper reweaving of the biological tapestry.

Quality-controlled tissue microdissection workflow is best supported by the creation of a tissue microdissection core laboratory or its equivalent. Metadata management is best when it is cleanly integrated into the workflow, and when these data automatically become analyzable along with specific molecular data generated downstream. This chapter discusses important considerations in establishing a high-quality tissue microdissection core laboratory and in establishing appropriate management of tissue microdissection metadata.

4.2 The Tissue Microdissection Core Laboratory: Organization and Basic Data Management

At the very least, a tissue microdissection core laboratory (TMCL) provides a basis for cost-sharing for access to expensive technology needed by many laboratories in a given institution or geographic area. At its best, a TMCL can provide a nexus for the morphologic know-how of an investigator to

Principles and Practice
Emmert-Buck et al. (Eds.):
Dissecting the Molecular Anatomy of Tissue

synergize with the dissecting know-how of the TMCL staff, improving the odds that a given dissection will yield material of the highest quality for downstream analysis.

Based on a survey of six current TMCLs including our own, most TMCLs have only bare bones support from their institutions, and this seems unlikely to change in the near future. Rather than advocate a pie-in-the-sky solution for laboratories with limited resources, we will take a highly practical approach to suggesting ways to improve your TMCL function:

1. Create a website. This can save a lot of time in explaining the technology you have available, and help potential users plan for experiments and costs associated. Examples of current TMCL websites are contained in Table 4.1 for reference.
2. Make sure that investigators know how to document dissections and to obtain images from dissections if needed. Having a set procedure for this will save an enormous amount of time if the investigator comes back in need of data for a publication and it is not available or not well organized. At the very least, encourage investigators to document the minimal list of items contained in Table 4.2.
3. If you have enough users, invite them for occasional seminars where they can present results they have obtained using microdissection, and/or develop an e-mail list to keep users informed. This type of interaction can stimulate new ideas and further scientific knowledge.
4. If you have enough users, time, and money, establish a database for managing your LCM laboratory and related tissue management. We have created software for this purpose, currently licensed and under development by BioFortis Inc. This is described below.
5. Get feedback from researchers on the results of their dissections. If there are problems, identify their cause and report this to other users of the facility. Use this feedback to establish quality control methods.

Table 4.1. A sample of current tissue microdissection core lab websites

Institution	Web address (Oct. 2003)	Comment
National Cancer Institute	*http://home.ncifcrf.gov/ccr/lop/Clinical/lcm/default.asp*	Ms. Virginia Espina
Medical College of Georgia	*http://www.mcg.edu/Core/Labs/lcm/Instrumentation.htm*	Dr. Jeff Lee
University of Alabama at Birmingham	*http://lcm.path.uab.edu/*	Dr. Isam Eltoum
University of Chicago	*http://lcm.bsd.uchicago.edu/*	Dr. Maria Tretiakova

Table 4.2. Minimal (M) and optimal (O) data needed for tracking microdissected tissue

Data type	M/O	Note
Subject unique identifier	M	Linked or unlinked to the name of the individual of origin as required for study, linked or unlinked to hospital record system as required by study
Tissue block unique identifier	O	
Tissue slide unique identifier	O	
Tissue slide type	O	
Tissue section thickness	M	
Tissue fixation type	M	Usually 4% buffered formaldehyde (AKA 10% neutral buffered formalin), 70% alcohol, frozen
Tissue fixation time	O	
Date/time of loss of perfusion of tissue	O	Time between loss of blood perfusion and time of complete freezing or fixation is likely to be a critical variable for some molecular studies. Time of exposure to fixative (if any), and processing parameters also can have important effects
Date/time of fixation/freezing	O	
Date/time of processing start	O	
Date/time of processing end	O	
Processing method (if any)	O	
Slide staining method	M	
Gross anatomic origin of tissue block or cells on slide	M	
Unique cell type(s) included in dissectate and method used to identify cells as such	M	
Annotated pre- and post- images of dissection	O	
Quality (integrity) of tissue based on standard histologic examination	O	
Ambient humidity at time of dissection	O	May have effects on activity of proteins in tissue during dissection and on ability to transfer tissue if using laser capture microdissection
Type of microdissection performed	M	Manual microdissection, laser capture microdissection, PALM-based direct slide catapulting, PALM-based membrane slides, Leica LMD membrane slide, etc.

Table 4.2 (continued)

Data type	M/O	Note
LCM cap (if any) unique identifier	O	
LCM pulses used (if applicable)	O	
Laser power used	O	
Dissected material microcentrifuge tube unique identifier	O	Needs to be linked to the specific slide used for each dissection
Notes	O	Recorded as needed for each microcentrifuge tube

4.3 Integrated Data Management for Tissue-Oriented Biological Research: The Labmatrix System

The Johns Hopkins PELICAN (Project to ELIminate lethal CANcer) laboratory developed an integrated multidimensional database for the management of tissue-oriented biological research that incorporates comprehensive daily laboratory workflow including tissue microdissection and continuous curation of data at the functional genomic level. This technology is licensed to BioFortis Inc. and has been named Labmatrix (Fig. 4.1, see also color illustration at the end of book).

In 1996, we established a translational research laboratory to perform detailed clinical-genomic studies in lethal metastatic prostate cancer. To integrate clinical data, sample tracking, and molecular data for this long-term project, Dr. Bova created the PELICAN Informatics laboratory. He employed a team of information technology professionals to create a prototype web-based secure database technology to allow data collection and review by the research team from various locations. From the outset, the technology was created to support "good data housekeeping" in any life science laboratory. The database model is research study subject-centric (human, mouse, or other species), and the point of view maintained is of the scientist bridging the chasm between phenotype and biomolecular data.

By 2001, demand from other laboratories to use the incomplete prototype system outstripped the ability to support the software properly, and BioFortis Inc. was formed to continue development of the technology under license from Johns Hopkins University. Labmatrix technology is now undergoing preliminary commercial testing in translational research laboratories.

The primary and support modules contained in Labmatrix are listed in Table 4.3. Labmatrix allows researchers to track study subjects' medical and phenotype data, biomaterials, a select set of "bench" laboratory and

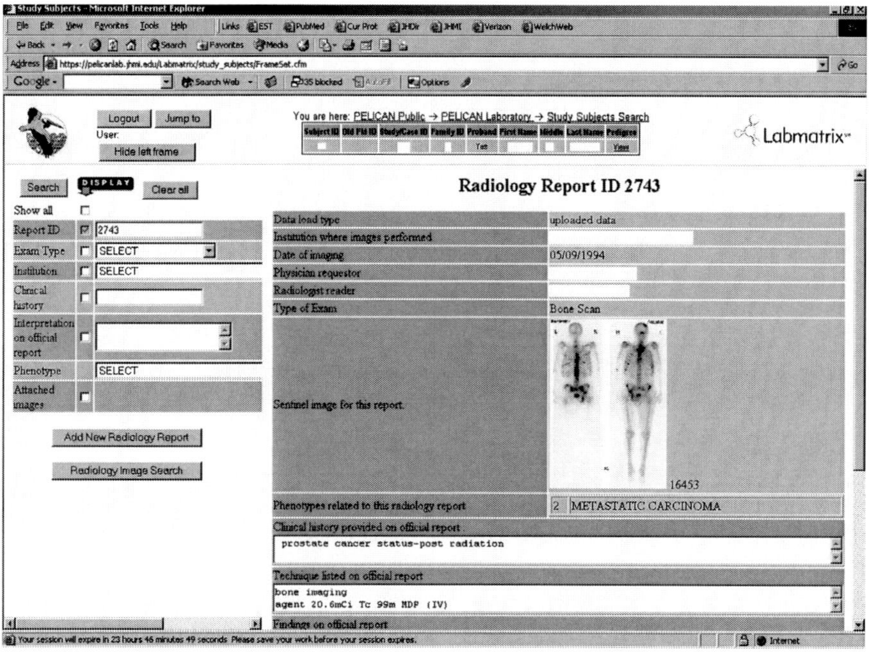

Fig. 4.1. Labmatrix example home page (individualized for each laboratory and user) (see color illustration at the end of the book)

Table 4.3. Labmatrix primary modules and support modules

Primary research modules	Laboratory support modules
Study subjects	Equipment and supplies
Biomaterials	Contacts
Protocols	Images
Scientific query builder	Configuration tools
	System support

genomic data and administrative data such as supplies and equipment. All of these data are gathered in the context of curated reference data (for example, a hierarchical list of human cell types). When completed, the platform will allow researchers to make queries, aggregate "locked" data for analysis, and allow the creation of figures and reports directly from the locked primary data. The result is a much more convenient, secure, and powerful way to manage and promote creativity in a single laboratory, or multiple laboratories working together.

4.3.1 Labmatrix Primary Modules

4.3.1.1 Study Subjects

In the study subjects module all types of clinical, family history, family relations, pedigree, and subject phenotype data are recorded and searched (Fig. 4.2, see also color illustration at end of book). Labmatrix users can link subjects and specific phenotypic data to disease terms directly from the UMLS (Unified Medical Language System), licensed from the National Library of Medicine. Users can collect, view and manipulate all types of clinical data and reports, including radiology images. Clients can access and track Institutional Review Board protocols, consent forms, authorizations, contacts, build custom surveys, and to-do lists for each subject, with appropriate standard data references (e.g., ancestry terms, relationship to proband terms, etc.) curated by BioFortis. Survey questions, for example, are drawn from a library of questions managed by BioFortis. In addition to complex password, role-based access, and user-customized Internet protocol (i.p.) address-based restrictions, confidentiality is supported by system features that are optional individual identifiers that are recorded in Labmatrix. These system features are available only in the study subjects part of the application.

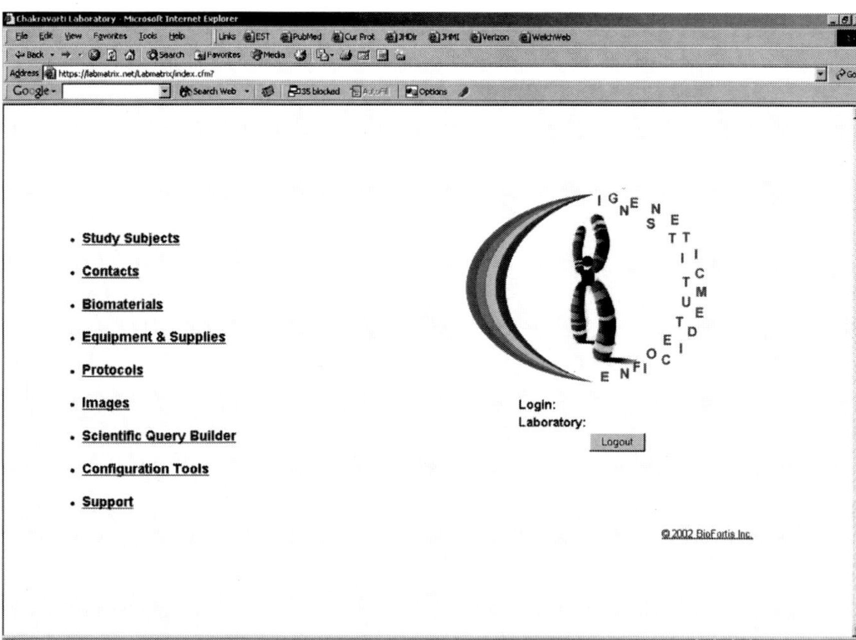

Fig. 4.2. Study subject module portion of radiology report and image (see color illustration at the end of the book)

4.3.1.2 Biomaterials

The biomaterials module is where critical information on tissue samples, tissue microarrays, body fluids, cell lines, DNA, RNA, and protein samples are managed in relation to the protocols used to create them and their physical location in refrigerators and freezers. Samples are organized according to their anatomic and cell-type source using a reference hierarchy provided by BioFortis, and maintained using bar-coding technology which is seamlessly integrated into the system. Currently, little or no tracking of the provenance of biomaterials is performed in the vast majority of bioscience laboratories, mainly because there is no convenient way to do so. This can create serious data integrity issues downstream, when for example, cDNA array (RNA expression) studies are based on tissues whose time between perfusion loss and freezing or fixation are unknown, and when the number of freeze-thaw cycles that a given sample has undergone are unknown, a critical issue in many proteomic studies.

4.3.1.3 Protocols

Protocols is a hierarchically organized workflow module designed to manage the day-to-day bench activities of the laboratory. It is an easy-to-use,

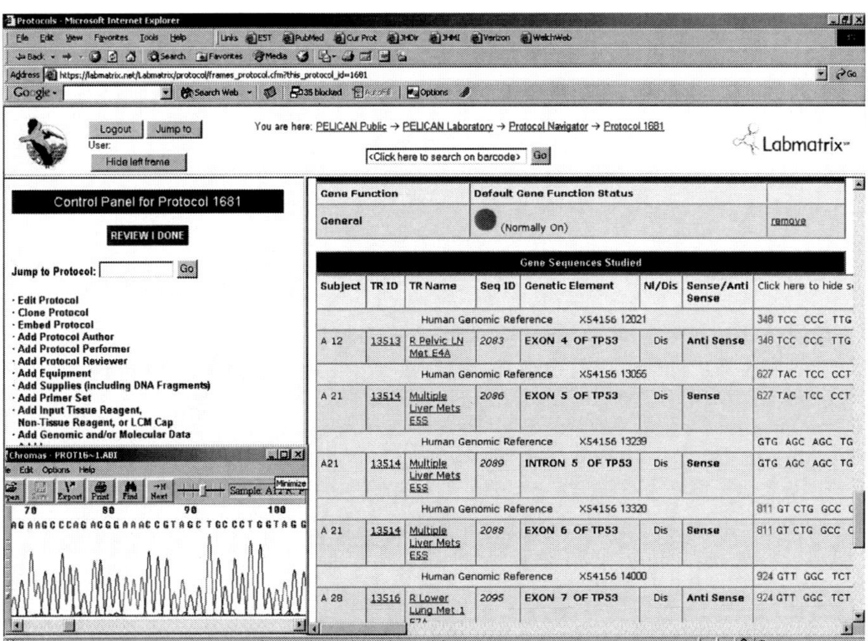

Fig. 4.3. Protocols module: portion of a genomic sequencing record (see color illustration at the end of the book)

powerful, real-time, flexible method for recording the input and output of specific work in the laboratory, such as the isolation of new DNA samples from tissues or body fluids, or the generation of sequence, comparative genomic hybridization, cDNA microarray, or immunostaining data. Protocols are simple to create through a user-friendly interface, and serve as a record of today's laboratory activity as well as a reference source for general methods used in the laboratory. Protocols provides the user with a convenient place to easily record methods and results (including images) accurately in real time, eliminating the need for time-consuming and error-prone rewriting of methods and results when a series of experiments is complete and ready for analysis and publication. Already in use in pilot laboratories, this system has improved the efficiency of preparing and submitting publications. An effort to develop the means to objectify and perhaps quantify Labmatrix influence on efficiency is one of the current aims during testing. Protocols also contains an integrated DNA fragment (PCR primer, etc.) management system, allowing clients to create and maintain references for DNA fragments to available public human genome data in real time (Fig. 4.3, see also color illustration at the end of book).

4.3.1.4 Workflow-Based Data Review: Functional Genomic Status Summarization Method

As currently configured, any user-created Labmatrix protocol wherein new study subject-based molecular data are obtained is required to go through a semi-automated review process. Users listed as reviewers (reviewer 1, reviewer 2, etc.) are notified on the Labmatrix^{S} logon screen with a link to the protocol in need of review. The review function proceeds through three stages: (1) final review of accuracy and completeness of data recorded in protocol, (2) automated or semi-automated review and reduction of the annotated molecular data. For example, in the case of genomic sequence, each experimentally derived study subject sequence is compared to a user-provided genomic reference (from public genome databases currently) and the user is prompted to annotate mutations using a standard mutation annotation method. (3) The third and final phase of review of molecular data-producing protocols is to place the newly obtained, annotated data in a novel "virtual" functional genomic context.

4.3.1.5 Scientific Query Builder

In the scientific query builder, researchers construct queries of the full dataset collected in their laboratory or included in third party sources, allowing them to aggregate data that they wish to preserve and export from Labmatrix for publication of individual result reports, full scientific manuscripts, or patent applications. Clients can save queries to be run at a later date, and can obtain both text-based and unique graphical displays of functional genomic data. The scientific query builder also contains a method for users to create and generate standard reports that can be saved and used on demand.

4.3.1.6 Equipment and Supplies/Contacts

The equipment and supplies module provides the ability to efficiently produce and maintain inventories of available equipment and supplies, facilitated by user-customizable lists of known equipment, supplies and reagents. It also allows laboratories to track costs, locations, expiration dates, product details and service contracts for individual items. Accurate inventories allow laboratory workers to set-up and complete new experiments rapidly and allow supply ordering to be conveniently managed by a single member of the laboratory. Inventories are also very useful at grant-writing time to allow accurate estimation of the costs of experiments.

4.3.1.7 Images/Configuration Tools

The images module provides a separate search engine for images in the various Labmatrix modules. This includes radiology images uploaded through study subjects, laser microdissection and other histology images recorded in biomaterials, and gel images recorded in protocols. The stored images can also be used in combination with other modules. For example, the survey builder in the study subjects module can be used in conjunction with the images module to create a review process, which allows multiple scientists to view and score or analyze many images from slides of various the biomaterials being studied. This enables collaboration, and promotes smooth workflow and efficiency of the laboratory process. The configuration tools module manages several key functions in Labmatrix. These include a facility manager, expense manager, data monitor, label printer manager, laboratory administrator, and user default settings.

4.3.1.8 Laboratory Administrator

The laboratory administrator submodule allows the laboratory's administrator to add new Labmatrix users within the laboratory, to administer conveniently role-based access to specific modules and submodules within the application, and to create and change the encryption key for the laboratory.

4.3.2 Additional Features of Labmatrix Prototype

4.3.2.1 Internal Proprietary and Public Reference Databases and Collaboration Features

Labmatrix internalized reference databases allow scientists to use the best available reference material in the process of their research, avoiding time-consuming external searching, and allowing critically important links to these references to be maintained as research continues. Because it is web-based, but provides tools for maintaining high security, scientists can provide controlled access to specific collaborators at will. One of the existing Labmatrix functions yet to be fully developed is to allow researchers to conveniently perform closely integrated collaborative research with other researchers and laboratories when desired, allowing researchers to bridge the significant domain knowledge gaps that exist in most current research environments.

4.3.2.2 Data Security Features and Regulatory Compliance

Researchers often need to keep data confidential during the discovery process, to protect patentability and for validation prior to publication. Labmatrix requires complex passwords for entry, and automatically inactivates usernames when more than five incorrect passwords are sequentially inserted (preventing successful dictionary attack). Labmatrix operates exclusively in secure sockets layer (SSL), so that all data transmissions are strongly (128-bit) encrypted, the current standard for web-based data transmission security. Labmatrix also provides a convenient method for researchers to limit access to their Labmatrix site through IP address restriction, and will do so in the future through the use of non-IP restricted security certificates. Privacy of data is also a critical consideration in research involving patient information and clinical records. Labmatrix provides several methods to allow researchers to conveniently adhere to published federal HIPAA (Health Insurance Portability and Accountability Act) rules and other requirements imposed by Institutional Review Boards and federal statutes covering human and animal subjects' research.

4.3.2.3 Integration with Other Scientific Tools and Technology

Labmatrix is a comprehensive "smart" container, workflow engine, and search engine which was intended to store primary data, and the results of analyses of primary data by external analytic tools. Based on sheer common sense, Labmatrix and other systems currently in development will likely provide major advances in science not currently possible using existing technology through vastly increased efficiency, increased ability to capture important molecular "exceptions to the rule" and unleashed scientific creativity. Whether this common sense is true needs to be proven in carefully designed studies of current workflow and database-enabled workflow in scientific laboratories.

Whether or not "individualized medicine" is necessary remains to be proven (perhaps generalities will continue to suffice), but only those medical centers that are able to integrate clinical, morphologic and molecular data from a specific patient with reference to populations of similar patients will be able to determine the value of such individualization. Establishment of Labmatrix or similar systems is thus a critical step in defining the future of medicine.

Acknowledgement. Special thanks are extended to the following directors of microdissection core laboratories for their responses to my informal survey of current tissue core labs operations: Ms. Virginia Espina (National Cancer Institute), Dr. Jeff Lee (Medical College of Georgia), Dr. Isam Eltoum (University of Alabama at Birmingham), Dr. Maria Tretiakova (University of Chicago), and Dr. Rob Edwards (University of California, Irvine).

Disclosure: under a licensing agreement between BioFortis, Inc. and the Johns Hopkins University. Dr. Bova and the University own BioFortis, Inc. stock, which is subject to certain restrictions under university policy. Dr. Bova also is a paid member of the BioFortis Scientific Advisory Board. The terms of this arrangement are being managed by the Johns Hopkins University in accordance with its conflict of interest policies.

II Methods of Tissue Microanalysis

5 Tissue Microarrays

Stephen M. Hewitt

5.1 Introduction

Pathology has acquired a high throughput platform with the advent of the tissue microarray (TMA). Correlating pathology with molecular biology is essential in biomedicine. Until the TMA was developed, most methodologies that used tissue were either labor-intensive, both in the performance of the assay as well as interpretation of the results, or "grind or bind" in approach. TMAs came about in their current form shortly after the advent of laser capture microdissection (LCM), and their uses are very closely tied.

Unfortunately, tissue microarrays were poorly named, and there has been extensive confusion about what they are as a result. A TMA refers to a slide of cores of tissue arranged in an organized grid. These cores of tissue are removed from *donor* blocks of tissue, and deposited in a *recipient* block. This recipient block is then sectioned to produce the TMA which is used in the final experiment (Kononen et al. 1998). The basic idea of a TMA does not greatly differ to the concept of Hector Battifora, in which tissue was formed into a *sausage* (Battifora 1986). However, the strict grid arrangement of the tissue as well as the significant increase in the number of tissues that can be placed in a single recipient block catapulted the utility of multiple tissue sections to a new level.

5.2 Construction of Tissue Microarrays

A complete guide to the construction and design of TMAs is beyond the scope of this chapter. Some general parameters will be reviewed, and there are numerous chapters and articles which provided significant detail in the construction and design of TMAs (Hoos and Cordon-Cardo 2001; Hewitt 2004). The majority of TMAs are constructed from paraffin-embedded tissue, however, very specialized TMAs are occasionally constructed from fro-

Principles and Practice
Emmert-Buck et al. (Eds.):
Dissecting the Molecular Anatomy of Tissue

zen tissue (Fejzo and Slamon 2001). One alternative to a frozen tissue array is the frozen protein array, in which cell lysates are placed in an array format, rather than whole tissue (Miyaji et al. 2002). The majority of paraffin-embedded tissue arrays are constructed from formalin-fixed tissue, although a number of researchers have used 70% ethanol as a fixative (see Chap. 4, this Vol.; Gillespie et al. 2002).

Most tissue arrays are constructed from human tissue, either archival specimens of tissues removed and processed for diagnostic evaluation or tissue collected specifically for research. Other organisms can be arrayed, although small rodent arrays can be difficult because of the size of the donor organs and tissues. Cell lines and tumor xenografts are being increasingly used in TMAs. The use of TMAs is not limited to cancer. However, tumors lend themselves readily to arraying by their very nature as a solid growth. The National Cancer Institute as well as numerous grantees have created tumor-specific and multi-tumor TMAs. Numerous companies offer TMAs of different malignancies. The most popular tumor systems are prostate and breast cancer. TMAs have been constructed for use in the teaching of normal anatomy, in studies of cystic fibrosis, and in aging models of the mouse and rat. In general, any system in which tissue is to be stained and compared to other tissues is amenable to the construction of a TMA.

Owing to the size of the cores of tissues placed on the array (0.6, 1.0, 1.5, or 2.0 mm in diameter), the ultimate density of the array, i.e., the total number of samples on the array, can vary from 30 (with 2.0-mm needles) to more than 1000 (with 0.6-mm needles), however, the practical density is around 500 cores per block. Although this may not reflect the exact number of samples on the array, it is possible for an entire study to be placed on a single slide. In the case of large epidemiologically based studies, a family of arrays may be required, which means increased throughput compared to the prospect of reviewing individual slides of each case for each immunohistochemical stain. Thus, a single TMA can provide sufficient information to confirm the findings of a small experiment or the results of experiments initially discovered by LCM or other high throughput methodologies.

5.3 Tissue Microarrays as a Validation Tool

Tissue microarrays are much like baseball trading cards; investigators can collect them and trade them with their colleagues. This has become one of their key benefits. If you wish to validate or compare a study for which a TMA was used, it is relatively easy to obtain slides to perform new experiments. Previously, institutions, as well as individual investigators, were hesitant to share material, not only because of ethical concerns and reviews, but also due to fear that the material might be exhausted. Because a TMA is a *biopsy* of a donor block, this issue has been ameliorated. Most

TMA blocks, depending on how they are sectioned, will produce between 50 and 300 high quality slides for analysis, far more than typically necessary for any single project. In addition, many TMA projects produce more than one recipient block, increasing the number of slides that will be produced. The enormous amount of data produced from one TMA or set of TMAs is a significant issue, similar to the datasets from microarrays. As a result, a data exchange platform has been created for TMAs (http://www.biomedcentral.com/1472–6947/3/5). This platform is designed to encode not only the information that went into the array and how it was constructed, but also the results of the experiments performed on the array – either at the level of the interpretation or as images of the results.

For many investigators, the issue is where can they find a TMA. As previously mentioned, there are commercial as well as government and institutional providers of TMAs. No single clearing house for TMAs currently exist. Of course, this is only an issue for the investigator who does not possess his own collection of material, or does not have a need to correlate it with other studies. The cost of arraying a collection of material varies. Some institutions have core facilities that will provide the service, and many academic pathology departments have arrayers, so that a collaboration can be easily fostered. For the individual investigator who is not a histopathologist, the challenges of the pathology inherent in constructing an array, as well as the cost of the instrument, do not make it practical to construct a TMA as a one-time project. However, an investigator, who has a collection of tissues sufficient to warrant arraying, probably has access to the expertise necessary to construct an array. It should be noted that there are commercial entities that will provide arraying services for a fee. Typically, these companies also offer commercial TMAs.

Because a tissue microarray is basically no different from a whole section of tissue, the ethical issues about obtaining the original material are no different. These have been summarized in Chapter 3. It should be noted that before the TMA is constructed, these issues must be resolved. The investigator who wishes to use the TMA needs to obtain only one approval to use the array and, depending on the source of the array as well as the clinical annotation, no approval may be required, e.g., TARP arrays and commercial arrays. Custom arrays or those constructed from specific protocols may require approval from the appropriate ethics oversight committee especially for access to clinical data.

Like LCM, the skills to fully use a TMA are based very much on the histopathology. Similar to LCM, what goes in determines the quality of what comes out. One significant difference is that much of the time-consuming pathology required for construction and interpretation of a TMA is spread out, and not as dependent on having a histopathologist available at all times. The quality of a TMA for diagnosis is determined by what goes into the array – the quality of the tissue used and the quality of the section of the material on the array. Whereas basic histology is not that difficult to learn, pathology can be very challenging, especially for complicated tumor systems, and the value of the pathologist cannot be over-emphasized. The advantage here is that the pathologist selects the tissue, at his/her leisure, the array is constructed, experiments are performed, and only at the time of interpretation, are the skills of the pathologist required again. Fortunately, most experiments with TMAs produce results that do not require review on a time-critical basis. To some, it sounds conceited that a histopathologist is required to review the results, and to some extent this is true, but the benefit of at minimum a consultation with a histopathologist cannot be emphasized enough. If the experiment is a simple immunohistochemical stain with a nuclear staining pattern, limited to the tumor cells only, then the histopathologist needs only to provide some guidance and to review the TMA to ensure that the staining was adequate. However, based on the experience of the author, the interpretation of some immunohistochemical stains is very difficult, and the review of a TMA has often completely altered the results of an experiment. This is especially true for experiments in which the distribution and intensity of expression must be interpreted. One especially challenging subject is the interpretation of in situ hybridization, which requires a level of expertise that most practicing pathologists lack.

5.4 Immunohistochemistry and Tissue Microarrays

The uses of a TMA are as varied as the tissue placed on a slide. The general rule is, if you can do it on a whole section, you can do it on a TMA. The primary use of TMA is immunohistochemistry, which accounts for approximately 80% of TMA experiments. The remaining 20% is split between in situ hybridization for mRNA transcripts, fluorescent in situ hybridization (FISH) for DNA and other uses from the mundane – traditional histochemical stains, to the exotic – infrared spectroscopy. Some time will be spent on the three main uses of TMAs and how they relate to LCM.

It is a strange and sad fact that TMAs at times have actually been the source material for some LCM experiments. This reflects the challenges that investigators face in obtaining sufficient samples with which to perform experiments. In the case of the original TARP arrays, produced at the National Cancer Institute, 75 different prostate tumors were arrayed on the TMA, along with 375 other tumors. These 75 prostate tumors represented far more cases than most investigators could obtain from any one source. They also come pre-reviewed by the pathologist, and annotated, so that the investigator must only dissect tumor cells from stroma, a relatively easy task compared to learning the complex histopathology of the prostate. This should be in no way be considered an endorsement of the approach, rather a sad commentary on the current state of affairs in obtaining tissue for biomedical research.

It is not surprising that immunohistochemistry is the primary use of TMAs, considering it is the most frequent application to embedded tissue after the H&E stain. It is also the easiest application, with numerous kits available from a variety of manufactures, and increasing automation of the process. Unfortunately, immunohistochemistry is frequently not well carried out, and an interpretation of the results is even more problematic. The failure of the methodology is frequently the absolute lack of controls and the very nature of the data it produces. There are reasonable negative controls that can be performed, but positive controls are limited to staining of tissue that has been previously characterized. Moreover, there is the difficulty of interpretation. Immunohistochemistry not only demonstrates whether a cell expresses a protein, but also where this expression is localized, within certain limits. This complexity is the inherent problem. Interpretation is not simply the presence or absence, or even the relative abundance of a band on a gel, but rather the deposition of a chromagen in a portion of a cell or extracellular structure that requires interpretation. This interpretation, at best, can be nominally described and quantified without applying sophisticated image analysis systems (discussed later).

Despite these negative aspects, immunohistochemistry is possibly the most powerful technique that can be applied to tissue by virtue of its capacity to localize the presence of a protein. Immunohistochemistry comes in many flavors and varying techniques, which are beyond the scope of this chapter. However, salient choices and their effects on the data will be discussed, especially as they relate to TMAs.

In a broad sense, immunohistochemistry is usually divided into two groups based on the method of detecting a signal – chromagenic and fluorescent. Chromagenic detection methods are the overwhelming choice: (1) they only require a standard microscope and (2) through the use of counterstains, interpretation of expression patterns is relatively easy. It is also possible to determine with ease which cells stain, the location of the protein in the cell, and rudimentary qualitative descriptions of abundance can be applied. The disadvantages of chromagenic detection methods include: (1) they are insensitive, (2) nonlinear in signal, and (3) have an extremely narrow window of signal strength (from no signal to maximum signal). Even with automation, these limitations cannot be overcome, however, there is no indication that chromagenic detection methods will be superseded in the near future.

A significant limitation of fluorescent detection methodologies in tissue is the auto-fluorescence encountered in formalin-fixed, paraffin-embedded tissues (Baschong et al. 2001). This auto-fluorescence has two sources: (1) the cross-linking of proteins by formalin, primarily at the disulfide bonds and (2) the presence of residual paraffin in the tissue section at the time of staining. Obviously, when working with frozen tissue, these problems are obviated. This is one of the primary justifications for the construction of a frozen tissue array. The formalin-induced auto-fluorescence is worst in tissues that contain numerous intermediate filaments. This problem can be partially overcome by using better quality band-pass filters in the microscope. Alternatively, the use of 70% ethanol will reduce this problem, as ethanol is not a cross-linking fixative. The removal of residual paraffin requires prolonged and repeated use of high quality (fresh) deparaffinization agents.

Fluorescent detection methods can be divided into two categories: those in which the antibody is directly labeled with a fluorescent tag (primary labels) and those in which a secondary antibody carrying the fluorescent tag are utilized. The important difference is the linearity of the signal – in the case of primary labeling, the signal should be linear over an extended range, however, sensitivity is compromised. Fluorescent secondary antibodies are more sensitive, but at best, linear in signal over a narrow range of fluorescence. Due to the reagents available today, the use of fluorescent secondary antibodies has overtaken the use of primary antibodies that have been tagged. This is due, in part, to the ready availability of commercial monoclonal antibodies and a wide choice of different fluorescent secondary antibodies. Using TMAs, the use of fluorescent detection systems is chiefly limited to studies that require well-defined (however limited) quantitation or evaluate more specific subcellular localization and/or colocalization of other proteins. The use of two or more labels in fluorescent microscopy is easily accomplished with the correct instrumentation, but remains a challenge in the chromagenic field. Some investigators have used the same scanners used for microarray analysis, however, the resolution of these scanners severely limits histological detail (Haedicke et al. 2003). It should be noted that fluorescent signals are typically better defined in cellular localization because of a lack of chromagen diffusion during its generation – fluorescence is limited to where the tag is located, not where a color reaction precipitates.

Fluorescent labeling is used for confocal microscopy (Baschong et al. 2001). Confocal microscopy, traditionally a methodology of cell biology, allows the precise localization of signals by using either limited apertures (spinning disk methods) or laser scanning to reduce the detection of scattered light. It is most frequently used to study the precise subcellular localization of a protein, the trafficking (translocation) of a protein, or the colocalization of proteins. This method typically produces images at precise focal planes with less depth of field than those accomplished in bright-field microscopy (0.5-μm step units compared to the whole thickness of a section in bright field), and then reconstructs these images in a “Z-stack”. Confocal microscopy is almost always performed with objectives of 40× or higher magnification, and generates numerous image files. In the past, confocal imaging was typically applied to cells in culture or cells in model organisms, and less in cancer tissue. The advent of the TMA has opened new horizons for confocal microscopy – offering the opportunity to study the subcellular localization of specific proteins in pathologic tissue at a high throughput level. Confocal microscopy has always been data-intensive, but

TMA has raised this to a new level and is now advancing microscopy and data acquisition in a fashion unimaginable 5 years ago. For example, a simple three-fluorescent tagged slide (such as FITC, rhodamine, and DAPI) and a 0.5-µm step size will result in 30 images per single location per core. For the NCI60 cell line array, this will result in 1740 images for the 58 cell lines on the array!

The most significant reason why immunohistochemistry remains the most important application for TMAs is that it allows the investigator to examine the expression of the target protein in tissue. Many arrays are used for "target verification", i.e., the capacity to generalize a finding in a large number of samples (usually on a different sample set) at one time and to verify that the observation that is being tested is truly a significant process and not a function of sampling error. For TMAs this is most frequently the process of testing a "hit" from a microarray experiment, but may not be limited to microarrays, and may also be a result from a Western blot or other proteomic analysis. When the TMA is used to *target verify* a microarray result, a further consequence is that the finding is "moved up" the biological process, from the transcriptome to the proteome. It is by no means certain that a change at the transcription level will be reflected in a stoichiometric fashion at the level of the proteome. This is not to suggest that *target verification* of microarray results by immunohistochemistry is problematic. In fact, in some instances, there are significant advantages to the method. For instance, when the protein target undergoes some phosphorylation, the ability to *target verify* a microarray result at the level of functional activity adds significant biological value to the finding.

5.5 In Situ Hybridization and Tissue Microarrays

Tissue microarrays have led to a resurgence in the use of in situ hybridization for mRNAs in tissue (Geiszt et al. 2003). The original in situ methodologies utilized radioactively tagged nucleic acid probes to detect the presence of a complementary nucleic acid (DNA or RNA). The method offers great specificity as well as sensitivity, but at a significant cost. Alternative detection methods, including fluorescence, chromagenic and chemiluminescence-based detection methods, have been used, but lack the sensitivity of radioactive probes, which, when performed correctly, can detect and discriminate the expression of single copy genes. When applied to whole sections, in situ hybridization is too time-consuming, expensive, and technically challenging to allow broad generalizations of the results. In the commercial setting, a single in situ experiment on a single slide may cost more than US$ 1000, with limited scales of economy for multiple slides assayed at one time. With the advent of TMA, it was suddenly possible to probe hundreds of samples at one time. In addition, it is now possible to provide some level of signal quantitation when appropriate control probes are used for comparison. Today, with high resolution (5–25 μm) phosphorimagers, the results of a radioactive in situ hybridization can be quantitated directly, without the laborious task of scoring dark-field illumination or counting silver grains of a photo emulsion. Efforts are underway to develop better detection methods, including fluorescent and chromagenic means. The TMA has significantly rejuvenated the use of in situ hybridization, which had been overtaken by PCR methods over the last 15 years. Like immunohistochemistry, in situ hybridization is typically used for *target verification* of genes identified by microarray. This method is primarily used when the investigator wants to target verify the results of an experiment on a larger number of samples or because appropriate antibody probes to the gene product are not available. It is not uncommon to identify a gene by microarray analysis for which no antibody is available, let alone an antibody that works in paraffin-embedded tissue. If an investigator wishes to study the expression of a specific splice isoform of an mRNA, in situ hybridization may be the only reasonable approach. It should not be overlooked that in situ hybridization provides direct tissue localization of a signal, whereas in the "grind and amplify" approach of PCR, it remains uncertain which cells actually express the mRNA target.

Fluorescent in situ hybridization has taken on the conventional role of detecting DNA using fluorescent probes for the determination of chromosomal detectable aberrations. A very powerful technique in cancer research, FISH can be applied to the analysis of metaphase spreads for high resolution studies or to interphase nuclei, when sufficient nuclei can be imaged appropriately (Bubendorf et al. 1999). One of the key limitations of FISH is the cost - generating fluorescent probes is expensive. Being able to investigate numerous samples at one time, the TMA has become a popular platform for interphase FISH analysis. Since the detection method is fluorescence, the challenges that plague immunofluorescence can occur in FISH. Fluorescent in situ hybridization is frequently employed in breast cancer research. The performance of FISH on TMAs or any tissue in determining that the cells being imaged are, in fact, the cell type of interest depends on extensive training and experience. Because of the challenge of obtaining samples for FISH, most investigators are willing to use formalin-fixed samples, and have worked out protocols to minimize auto-fluorescence. Because the target of interest is DNA, and not protein, auto-fluorescence from cross-linked proteins can be more easily controlled, as the samples are typically digested with proteases.

5.6 Image Capture and Data Analysis

The development of TMA has become a driving force in automated image capture and analysis. Although automated image capture and analysis has been slowly advancing over the last decade, they have not found wide application. Pathologists are routinely overwhelmed with interpreting hundreds of slides, and the limitations of manual interpretation, as previously described, negatively impact the quality of research that can be performed. Fortunately, these limitations coincide with recent advances in digital photomicroscopy, computer power, and the recognition of bioinformatics as a science of its own. Currently, progress is being made in both brightfield and fluorescent imaging as high-speed, high-throughput processes. Now that large datasets are available, new analytic tools are essential. Without imaging and image analysis, chromagenic immunohistochemistry produces only nominal data, for which only limited statistical analysis is possible. Worse yet, the data is difficult to reproduce because of the subjective interpretation by pathologist. The pathologist excels at pattern recognition, but is poor at quantification. Now with automated image capture it is possible to develop algorithms that can discriminate and quantify the stain patterns of different markers in different tissues. Although a computer can quantify the intensity of a stain, or the number of positively staining nuclei on a slide, more effectively, the software remains limited. Most software lacks the chief skills of the pathologist, pattern recognition and interpretation. An excellent example is the understanding and quantification of the expression pattern and levels of certain proteins in renal cancer. The clear cell carcinoma of the kidney has tumor cells that are rich in lipid and glycoprotein, typically with a large vacuole, pushing the cytoplasm to the periphery of the cell. In contrast, papillary carcinoma of the kidney has tumor cells with more typical volumes of cytoplasm without vacuoles. It is practically impossible to quantitate an immunohistochemical stain for a protein between these two tumor cell types using automated image analysis.

Even with appropriate quantitative imaging tools, analyzing the data remains a subject of debate. Cluster analysis used for microarrays is inappropriate, as the choice of antibodies applied to a TMA is determined by the researcher. As a result, the study is completely supervised, not unsupervised, as required for unsupervised cluster analysis. Principal component analysis can provide useful visualization of data. The use of traditional statistical analysis including correlations between different markers, or more importantly, outcomes, remains the method of choice. Traditional statistics do not model complicated biological relationships, such as pathways, adequately. They are unable to incorporate signal amplitude into the models. Only with the large datasets that are now possible with TMAs will we be able to develop the new analytic tools.

Hopefully, this chapter elucidates some of the uses and challenges of using TMAs in research. The TMA is a research tool which increases the throughput of samples that can be analyzed. When properly designed and used, the limitations of the TMA are significantly outweighed by the benefits of being able to condense hundreds of analyses onto one slide.

References

Baschong W, Suetterlin R, Laeng RH (2001) Control of autofluorescence of archival formaldehyde-fixed, paraffin-embedded tissue in confocal laser scanning microscopy (CLSM). J Histochem Cytochem 49:1565–1572

Battifora H (1986) The multitumor (sausage) tissue block: novel method for immunohistochemical antibody testing. Lab Invest 55:244–248

Bubendorf L, Kononen J, Koivisto P, Schraml P, Moch H, Gasser TC, Willi N, Mihatsch MJ, Sauter G, Kallioniemi O-P (1999) Survey of gene amplifications during prostate cancer progression by high-throughout fluorescence in situ hybridization on tissue microarrays. Cancer Res 59:803–806

Fejzo MS, Slamon DJ (2001) Frozen tumor tissue microarray technology for analysis of tumor RNA, DNA, and proteins. Am J Pathol 159:1645–1650

Geiszt M, Lekstrom K, Brenner S, Hewitt SM, Dana R, Malech HL, Leto TL (2003) NAD(P)H Oxidase 1, a product of differentiated colon epithelial cells, can partially replace $gp91^{phox}$ in the regulated production of superoxide by phagocytes. J Immunol 171:299–306

Gillespie JW, Best CJ, Bichsel VE, Cole KA, Greenhut SF, Hewitt SM, Ahram M, Gathright YB, Merino MJ, Strausberg RL, Epstein JI, Hamilton SR, Gannot G, Baibakova GV, Calvert VS, Flaig MJ, Chuaqui RF, Herring JC, Pfeifer J, Petricoin EF, Linehan WM, Duray PH, Bova GS, Emmert-Buck MR (2002) Evaluation of non-formalin tissue fixation for molecular profiling studies. Am J Pathol 160:449–457

Haedicke W, Popper HH, Buck CR, Zatloukal K (2003) Automated evaluation and normalization of immunohistochemistry on tissue microarrays with a DNA microarray scanner. BioTechniques 35:164–168

Hewitt SM (2004) In: Fung E (ed) Protein microarrays: methods and protocols. Design, construction and use tissue microarrays. Humana Press, Totowa, New Jersey

Hoos A, Cordon-Cardo C (2001) Tissue microarray profiling of cancer specimens and cell lines: opportunities and limitations. Clin Invest 81:1331–1338

Kononen J, Bubendorf L, Kallioniemi A, Barlund M, Schraml P, Leighton S, Torhirst J, Mihatsch MJ, Sauter G, Kallioniemi O-P (1998) Tissue microarrays for high-throughput molecular profiling of tumor specimens. Nat Med 4:844–847

Miyaji T, Hewitt SM, Liotta LA, Star RA (2002) Cryoarray, a new platform for proteomics of recombinant and native proteins. Proteomics 2:1489–1493

III Molecular Analysis of Microdissected Samples

6 Methods of Microdissection and New Technologies

Michael A. Tangrea, John W. Gillespie, Michael J. Flaig, Gallya Gannot, Vladimir Knezevic, Michael R. Emmert-Buck and Rodrigo F. Chuaqui

6.1 Historical Perspective

The analysis of tissue samples is challenging due in large part to the complexities of normal and pathological histological structures. This is a major factor to consider when applying molecular techniques to patient or animal model specimens. For example, human prostate cancer tissue sections frequently contain, in addition to prostate cancer cells (which may be identified in one or multiple foci), pre-invasive cells [prostatic intraepithelial neoplasia (PIN) foci], epithelial hyperplastic foci, normal glands, stroma, inflammatory infiltrate, and nerves. In fact, it is not uncommon that prostate cancer foci constitute only a small fraction of the total number of cells contained in one tissue section. Therefore, it is essential to apply methods that allow for the procurement of pure or near pure cell populations from complex tissue sections for molecular analysis. Tissue microdissection provides this opportunity.

The impact of tissue microdissection on the analysis of tissue specimens is particularly important in assays where the purity of cells is critical, such as the analysis of genomic deletions that occur specifically in tumor cells, since contaminating normal epithelium or nontumor cells may mask the deletion. Similarly, mRNA- or proteomic-based analyses may be compromised if the sample under study does not contain a relatively pure population of the cell type of interest. The development of polymerase chain reaction (PCR) technology in the 1980s led to an explosion in the number of studies identifying new deletion loci in tumors (Radford et al. 1993; Emmert-Buck et al. 1995; Koreth et al. 1995; Speiser et al. 1996; Thiberville et al. 1995; Zhuang et al. 1995b; Fuji et al. 1996; Chuaqui et al. 1997a). Gene expression studies, such as those based on cDNA library analysis, also benefited significantly from microdissection, since a direct comparison from each cell compartment in the progression of diseases, such as normal, premalignant and malignant cells could be performed. For example, the construction of a cDNA library from a microdissected PIN lesion showed a

Principles and Practice
Emmert-Buck et al. (Eds.):
Dissecting the Molecular Anatomy of Tissue

significant increase in the identification of novel ESTs compared to conventional libraries constructed from whole tissue blocks (Krizman et al. 1996). The purity of the PIN cell population allowed for the cloning of transcripts that are usually masked by more abundant cell populations, such as stromal, inflammatory or normal cells. More recently, gene expression analyses have been significantly extended using various microarray platforms (see Chap. 9, this Vol., for additional examples). Similarly, proteomic-based efforts have been aided by tissue microdissection, as described by Geho and colleagues in Chapter 10.

The field of microdissection has evolved through several stages. Initially, the trimming of tissue blocks to avoid unwanted areas, or the scraping of tissue in a desired area was utilized. However, the success of the gross dissection efforts was limited by the histology of the tissue under study, such as the pattern of tumor invasion. Good results were obtained from homogeneous tumors, such as lymphomas, or carcinomas invading as solid sheets of cells, such as those frequently seen in squamous cell carcinomas, melanomas, mesotheliomas or sarcomas. Poorer results were obtained in the case of more complex invasion patterns, such as lobular breast carcinomas, inflammatory tumors, or adenocarcinomas in general, where the tumor is composed of individual glands formed by a one- to two-cell layer and lumen, admixed with reactive stromal and other host cells.

An alternative approach for selective tissue analysis is to eliminate undesired areas, and subsequently, to scrape the entire remaining tissue, including the unaffected areas of interest. The destruction of undesired areas can be performed by either directly exposing the tissue to a laser (obliteration; Hadano et al. 1991), or by irradiating the tissue with ultraviolet light, a technique coined SURF (selective ultraviolet radiation fractionation; Shibata et al. 1992). In the most commonly utilized approach, protective ink is applied with a microdotter to the desired areas. The UV light irradiates the entire histological field and generates a permanent cross-link within the unprotected (noninked) areas. When the whole tissue is subsequently recovered, only the protected areas provide nondamaged, analyzable DNA.

Finer, positive selection microdissections were performed using a small gauge needle to scrape the desired areas under microscopic visualization (Radford et al. 1993; Emmert-Buck et al. 1994; Koreth et al. 1995; Speiser et al. 1996; Zhuang et al. 1995b; Chuaqui et al. 1997b; Quezado et al. 1999). These manual approaches provided acceptable purity in a significant proportion of the cases, frequently showing absence of contamination, even after PCR amplification of nucleic acids, as described by Uribe and Wistuba (Chap. 8, this Vol.). However, the method was slow, tedious and totally dependent on the operator's dexterity. Moreover, precision was not optimal, which frequently limited the studies to those cases where the infiltrative pattern corresponded to easier dissections (Rubin et al. 2002). In addition, the desired areas were destroyed as they were procured, thus a quality assurance image of the recovered cells could not be recorded.

Another microdissection advancement was the development of hydraulic manipulators, which enabled a more precise type of needle- or probe-based dissection (Going et al. 1996; Lee et al. 1998; Shegal et al. 1998; Wistuba et al. 2001; Obiakor et al. 2002). In most cases, a fine tipped Pasteur pipette was prepared under heat and used for scraping instead of an actual needle. The tip was placed on a movable stage that could be manipulated in the three axes (x, y and z). The method was precise and allowed, for example, a single layer lining from normal gallbladder to be procured (Wistuba et al. 2001), and single lymphocytes from rabbit spleen germinal centers to be obtained (Obiakor et al. 2002). However, the method was laborious and slow, features that are particularly problematic when a large number of cells are needed for the downstream analysis.

6.2 Laser-Based Dissection Technologies

At present, most tissue microdissection studies are performed using one of the commercially available laser-based methods. Provided below is a brief overview of each of these technologies. Additional information is available from the company websites that are provided, as well as in the cited references.

6.2.1 Laser Capture Microdissection

Laser capture microdissection (LCM) (Emmert-Buck et al., 1996; Bonner et al. 1997) was invented at the National Institutes of Health (NIH) and developed by Arcturus Engineering, Inc. (www.arctur.com) through a cooperative research and development agreement. The system utilizes a near-infrared gallium arsenide laser diode integrated into an inverted compound microscope with alignment into the optics subsystem. The principle of LCM is shown schematically in Fig. 6.1 A. The cap, which has a thin layer of thermoplastic polymer film containing an infrared-absorbing dye, is placed on the surface of a routinely prepared tissue section on a glass slide. The investigator examines the tissue section microscopically and activates the laser when the desired cells underlie the target. The activation causes the film to focally melt, binding very specifically to the subjacent cells. Procurement of the cells of interest from the tissue occurs by lifting the cap. To accommodate the dissection of areas as small as single cells to groups of cells, the laser has three settings for the beam diameter including 7, 15, and 30 μm. The LCM system software allows for the adjustment of the laser by varying the power (in the range of milliwatts) and the timing (microseconds to milliseconds) of the pulses. The software also utilizes image capture to document a tissue roadmap image, image of the tissue before and after microdissection, and cap image of captured cells. These are criti-

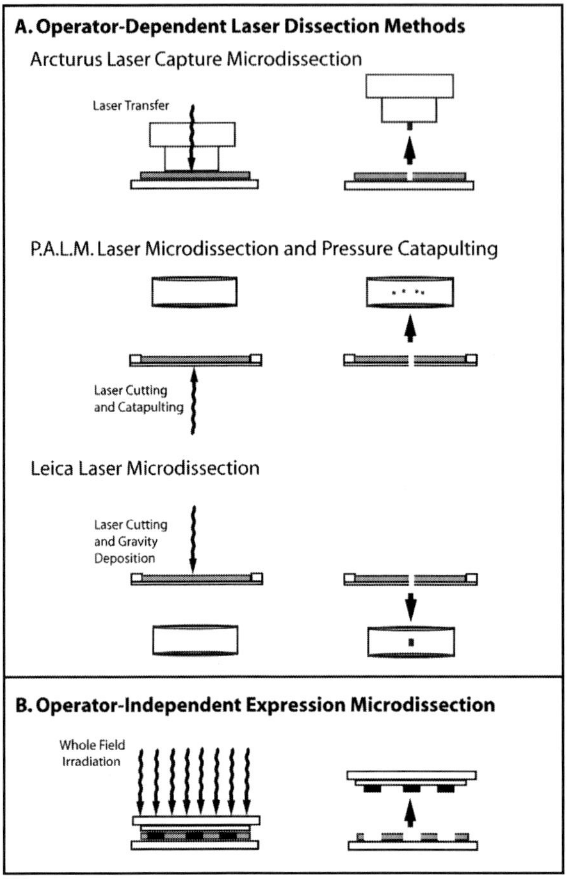

Fig. 6.1. Methods of laser microdissection. **A** In operator-dependent methods, the desired tissue areas are first identified by morphology and then targeted with the laser. In the Arcturus laser capture microdissection system the cells are transferred on a film, in the PALM method they are catapulted and in the Leica system they are cut and deposited by gravity. **B** In the operator-independent expression microdissection approach, the whole tissue field is irradiated after immunohistochemical staining. Only the stained cells are transferred to a film

cal in maintaining an accurate record of each dissection and correlating histopathology with subsequent molecular results.

The advantages of the LCM system are its simplicity, reliability, and speed. Unlike manual microdissection, the technique is easily mastered. Since the system uses an infrared laser instead of an ultraviolet laser, there is minimal damage to biomolecules. Even though the membrane is activated at 90 °C, the tissue experiences only a brief thermal change as the heat generated from the membrane is rapidly dissipated. The cells are transferred to the polymer film in one step and then proceed directly to analysis. LCM can be used for dissecting cells from almost all tissue types, including those that have been snap-frozen, or fixed and embedded, as well as cytological samples. The protocol for the preparation of slides for LCM is described in the Protocols section (Chap. 11.2.1). The dissected cells can be used for analysis of DNA, RNA, or protein as described in Chapter 11.4.

While early generations of LCM (PixCell) have been useful for a wide variety of applications, the latest generation, AutoPix, has several new features to provide a higher throughput platform for molecular profiling. It combines robotics and optical scanning software for automated microdissection of selected cells. The AutoPix has an enclosed configuration; therefore, there are no ocular lenses. Instead, it utilizes a phase alternate line (PAL) format color camera to permit visualization of the slide as a "roadmap image" for determining the target area for microdissection. The enhancements of the automated system include three-slide capacity, quantitation of the area of microdissected tissue, diameter measurement of the wetted polymer spot, and cell recognition software. The AutoPix imaging software creates index-matched, stitched images of the tissue that are utilized for cell selection, allowing more precise targeting of selected cells. Enhancement of the tissue image, coupled with optical scanning software, provides more accurate identification of cellular morphology during cell selection as compared to manual systems. Software tools allow the visualization of the projected laser pulse in relation to the tissue, making it possible to discern the exact amount of tissue that will be microdissected with each laser pulse. In order to precisely dissect a single cell, cell layers, or tissue areas, annotation software allows corresponding single point, line, or polygon options for dissection, respectively. Algorithm-based, cell image recognition software on the AutoPix platform enhances the automated features of this instrumentation. The algorithm is based on texture, morphology, size, color and contrast of the tissue, permitting automated cell selection in addition to automated microdissection. Similar to earlier generations of LCM, documentation of microdissected tissues is achieved through image capture.

6.2.2 PALM Laser Microdissection and Pressure Catapulting Technology

Originally dating back to 1998, the PALM MicroLaser System provides a different approach to laser microdissection (http://www.palm-microlaser.com). Based on the laser microdissection and pressure catapulting (LMPC) technology, the system provides a noncontact method for precise dissections of tissue samples. As an initial step, the tissue section is mounted on a 1.35 μm-thin polyethylene membrane slide (Schutze and Lahr 1998). The investigator then examines the tissue microscopically and digitally defines the tissue areas for dissection. The pulsed UV laser automatically performs a perimeter cut of the desired area. The laser is then used to produce a high phototonic pressure force enabling the ejection of the selected sample from the plane of the tissue. Through the generated force, the dissected area is catapulted into the cap of a microcentrifuge tube for downstream molecular analysis.

The LMPC system can be used with most routinely prepared cell and tissue samples, including living cell cultures (Stich et al. 2003). As reported by Schutze et al., the LMPC technology can dissect an area as small as 1–15 μm in diameter (Schutze et al. 1998), and can be used to dissect tissue for most standard molecular analysis techniques. Recent molecular studies have used the PALM system to isolate and analyze DNA and mRNA from melanoma cell lines, colon adenocarcinoma, rat lung and mouse liver tissue (Bersen et al. 1998; Fink et al. 1998; Schutze and Lahr 1998; Schutze et al. 1998; Burgemeister et al. 2003).

6.2.3 Leica AS LMD

Utilizing a nitrogen UV laser, the Leica AS LMD offers another alternative approach to laser microdissection (http://www.light-microscopy.com). Similar to the PALM system, the Leica AS LMD requires the investigator to microscopically examine the tissue prior to dissection. By integrating the microscope-viewable area with a digital camera image, the user manually defines the area of interest on the digital image and then allows the fully automated laser to dissect the appropriate area. The LMD system requires the tissue section to be directly mounted on a specialized frame foil slide. This specialized slide consists of a metal frame and a 1.4-μm polytarthalene (PET) membrane mounting area. The automated pulsed UV laser cuts the thin plastic film platform that supports the tissue section. The dissected tissue area falls by force of gravity into a tube for downstream molecular analysis, thus no mechanical or physical forces are needed for the dissection, resulting in contamination-free samples.

Possible sample sources include, but are not limited to standard histological sections and cultured cells. Through the use of this technique, it is possible to dissect an area as minute as 4–5 μm in diameter, thus allowing for highly precise cutting. Any number of downstream molecular techniques can then be applied to dissected tissue areas, including DNA, RNA and protein gel electrophoresis and array studies (Burbach et al. 2003). An apparatus using the same general principle is also produced by Bio-Rad Laboratories under the name of the Clonis system (http://www.bio-rad.com).

6.3 Novel Techniques

The complex milieu that makes up the molecular anatomy of tissue is one that is quite challenging to understand. With the advancement of laser-based microdissection techniques (Emmert-Buck et al. 1996; Bonner et al.

1997) and the recently developed high-throughput gene expression array (Fearon et al. 1987; Shibata et al. 1992; Radford et al. 1993; Emmert-Buck et al. 1994; Schena et al. 1995; Zhuang et al. 1995a; Chee et al. 1996; DiRisi et al. 1996; Going and Lamb 1996; Lockhart et al. 1996; Moskaluk and Kern 1997, Schena et al. 1998; Celis et al. 2000) and proteomic techniques (Emmert-Buck et al. 1996, 2000; Becker et al. 1997; Bohm et al. 1997; Anderson and Anderson 1998; Schutze and Lahr 1998), technology now makes it possible to separate normal and diseased cells and to analyze simultaneously thousands of mRNA transcripts and proteins from the specific cell populations. Typically, these efforts provide a novel insight into the molecular makeup of disease states (Zhang et al. 1997; Golub et al. 1999; Klose 1999; Alizadeh et al. 2000; Perou et al. 2000; Strausberg et al. 2000; Hedenfalk et al. 2001; Kurachi et al. 2001). Although these studies provide valuable information, they represent only the first experimental step. In the second stage, it is necessary to independently confirm and quantitatively measure the expression level of each of the genes of interest. At present, this represents a significant challenge in terms of time and effort due to the often limited amount of biological sample available, particularly in the case of clinical specimens. In order to optimize the use of preserved tissue specimens and obtain information from archived material, we are currently developing two new techniques, expression microdissection (xMD) and layered expression scanning (LES). These novel techniques enable the researcher to maximize the information gained from small, precious samples in hopes of obtaining a more accurate picture of the molecular profile of specific cell types. Both xMD and LES help to increase the output of gene expression measurements from a biological sample and can be directly applied to the analysis of human tissues, which is particularly valuable when performing follow-up studies of candidate genes. Though not directly related, xMD and LES in combination could be used to maximize the information gained from a tissue sample.

6.4 Expression Microdissection

For tissue microdissection to keep pace with the needs of today's investigators, it is important that faster and simpler approaches are contemplated and developed. Expression microdissection is a technique newly developed at the National Cancer Institute, Laboratory of Pathology that fulfills these requirements through the exploitation of the antibody–antigen relationship found in nature. The technique uses a targeting probe, such as an antibody or nucleic acid oligomer, to define the area of dissection, removing the "human element" from the process. xMD physically procures target cells using standard immunohistochemical (IHC) staining procedures, thermoplastic film and a light source (Fig. 6.2). Cells of interest are targeted using a specific antibody that focally deposits an absorptive stain. The entire his-

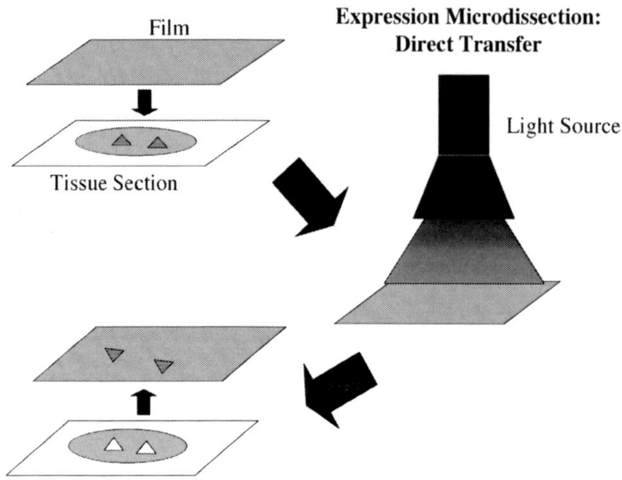

Fig. 6.2. Schematic illustrating expression microdissection (xMD)

tological field is then irradiated causing film activation selectively above cells that have been targeted. The transfer film is removed and the procured cells are ready for molecular analysis. xMD offers several advantages over current methods, including: (1) increased dissection efficiency (on the order of 4000 cells/s), (2) increased dissection precision (subcellular, such as nuclei), (3) removal of variance among individual operators, permitting standardization of the dissection process, (4) elimination of the need for a histopathologist, allowing the method to be carried out by scientists and technicians in a laboratory, (5) rapid procurement of specific cells based on their phenotypic profile, irrespective of their geographic location (clustered or scattered), and (6) elimination of difficulty in cell targeting due to the poor image quality of noncover-slipped, nonindex-matched histology sections utilized in current dissection systems (no visualization is required for xMD).

The physical principles underlying xMD have been thoroughly assessed, including laser dosimetry (wavelength, pulse time and number, power), transfer film characteristics, cellular stains, and heat sink effects of tissue section support surfaces. An optimized protocol has been developed to maximize film activation and adherence to the underlying substrate. xMD

Fig. 6.3. Procurement of specific cell types using xMD. **a** Procurement of an entire field of prostate epithelium using anti-cytokeratin as a targeting antibody. **b–d** Dissection of prostate epithelium stained for prostate-specific antigen (PSA), before (**b**) and after xMD (**c**). Cells recovered onto transfer film (**d**). **e** Basal epithelial cells (*arrows*) remaining after procurement of PSA positive secretory cells. **f** Retrieval of basal epithelial cells

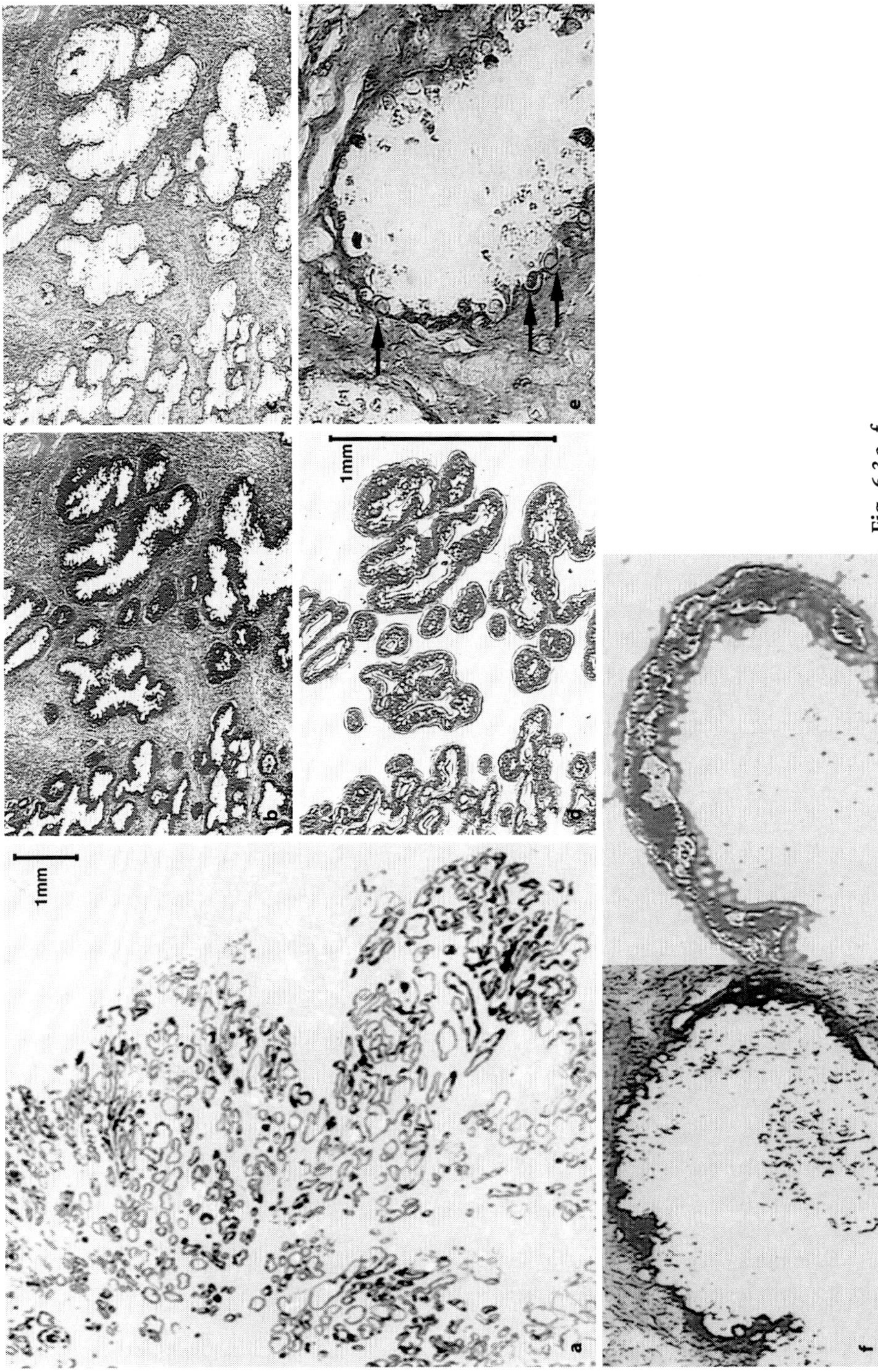

Fig. 6.3 a–f

has been applied to varying fixative processed histological sections, including snap-frozen tissue, formalin-fixed, paraffin-embedded specimens, and ethanol-fixed, paraffin-embedded samples, to evaluate the consistency of dissection. In each case where cellular staining was observed, xMD was successful in retrieving the labeled cells. Importantly, no film activation occurred in nontargeted regions of the tissue. Several examples of cell procurement by xMD are illustrated in Fig. 6.3. Figure 6.3 A demonstrates recovery of prostate glands from a human tissue section onto transfer film using cytokeratin as a targeting antibody. The epithelium from a 10-mm^2 area (approximately 100,000 cells) was precisely dissected in less than 200 s (rate $\cong$ 500 cells/s) using a prototype scanning laser system. Figure 6.3 B, C shows a section of prostate epithelium stained with an antibody against prostate-specific antigen (PSA), before (B) and after (C) xMD. The procured cells are displayed on the transfer film in Fig. 6.3 D.

Since activation of the film is specifically localized to the region of dye deposition, xMD is extremely precise. Close inspection of the dissected glands in Fig. 6.3 indicates that only the inner PSA positive secretory cells were removed. The surrounding, PSA-negative basal stem cell layer (counterstained with hematoxylin and eosin) was not procured (Fig. 6.3 E). The stem cells can then be subsequently dissected using either standard LCM, or with a second round of xMD using an antibody specific for basal cells, such as antibodies against high molecular weight cytokeratin (34betaE12 and p63; Fig. 6.3 F; Zhou et al. 2003). Determining the role(s) that secretory and basal stem cells play in normal prostate physiology and disease is of great interest; however, until now it has been extremely difficult, if not impossible, to individually recover enough of these cells for analyses. The wide range of antibodies available for tissue staining offers many possibilities for xMD. For example, studies of normal prostate and prostate cancer could be extended by using commercially available antibodies that target tumor cells, normal epithelial cells, neovessels, proliferating cells, apoptotic cells, infected cells, inflammatory cells, stromal cells, specific pathway-activated cells, p53 mutation-positive cells, or cells in a specific phase of the cell cycle.

Our experience to date indicates that xMD is capable of dissecting target structures smaller than a single cell. This is illustrated in Fig. 6.4 A–E. Figure 6.4 A shows a histological section of normal mouse colon stained for proliferating nuclear cell antigen (PCNA), which stains nuclei in the lower segment of colonic crypts. Superficial epithelium and upper segments of the glands are PCNA negative. Figure 6.4 B demonstrates successful procurement of PCNA immunostained nuclei. Similarly, dissection of cytomegalovirus (CMV) positive nuclei from infected human lung is seen in Fig. 6.4 C–E. A standard hematoxylin and eosin stained section is shown in panel C with characteristic nuclear viral inclusions (arrowheads). CMV immunopositive nuclei are evident in Fig. 6.4 D, and the recovered nuclei after xMD are shown on the transfer film in Fig. 6.4 E. In both the PCNA and

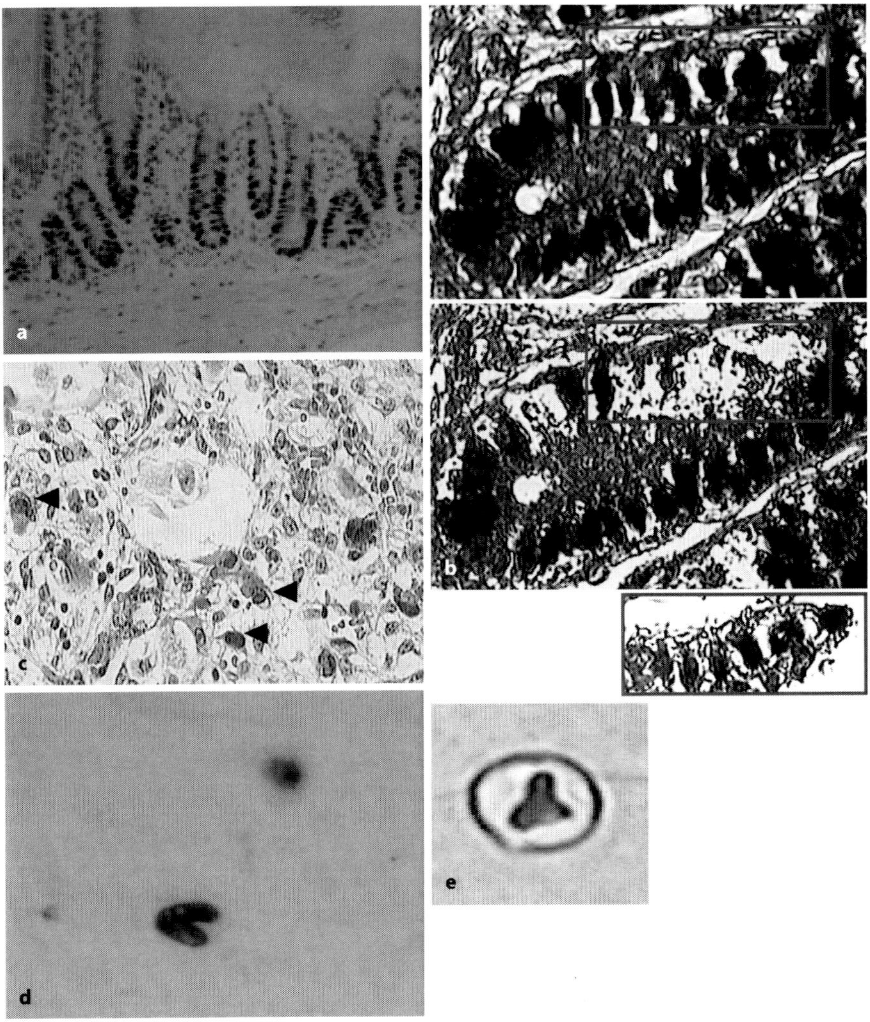

Fig. 6.4. Procurement of subcellular components using xMD. **a** Mouse colon section showing proliferating nuclear cell antigen (PCNA) positive nuclei of cells in lower segment of crypts. **b** Procurement of PCNA positive nuclei. **c–e** Dissection of cytomegalovirus (CMV) infected nuclei from infected human lung sections

CMV targeting experiments, three advantages of xMD are demonstrated: (1) precise subcellular dissection of nuclei, (2) expression-targeted dissection of a specific subset of nuclei, and (3) operator-independent dissection.

The biomolecule integrity after xMD treatment has been thoroughly evaluated using several common laboratory methods to analyze DNA, mRNA and protein quality individually. Similar to standard LCM, the cells are exposed to a brief thermal transient (5–10 ms) during film activation, thus it was important to determine whether the biomolecules were damaged during the process. Analyses of DNA, protein and mRNA are shown in Fig. 6.5 A–G. Figure 6.5 A demonstrates polymerase chain amplification (PCR) of DNA at three sequence tagged sites (STS). In this example, xMD used cytokeratin-based targeting of human prostate epithelium. Allelotyping of the sample using microsatellite marker D8S1786 was also performed (Fig. 6.5 B). Amplification of glyceraldehyde 3-phosphate dehydrogenase (GAPDH) from DNA recovered from PCNA-positive nuclei is illustrated in

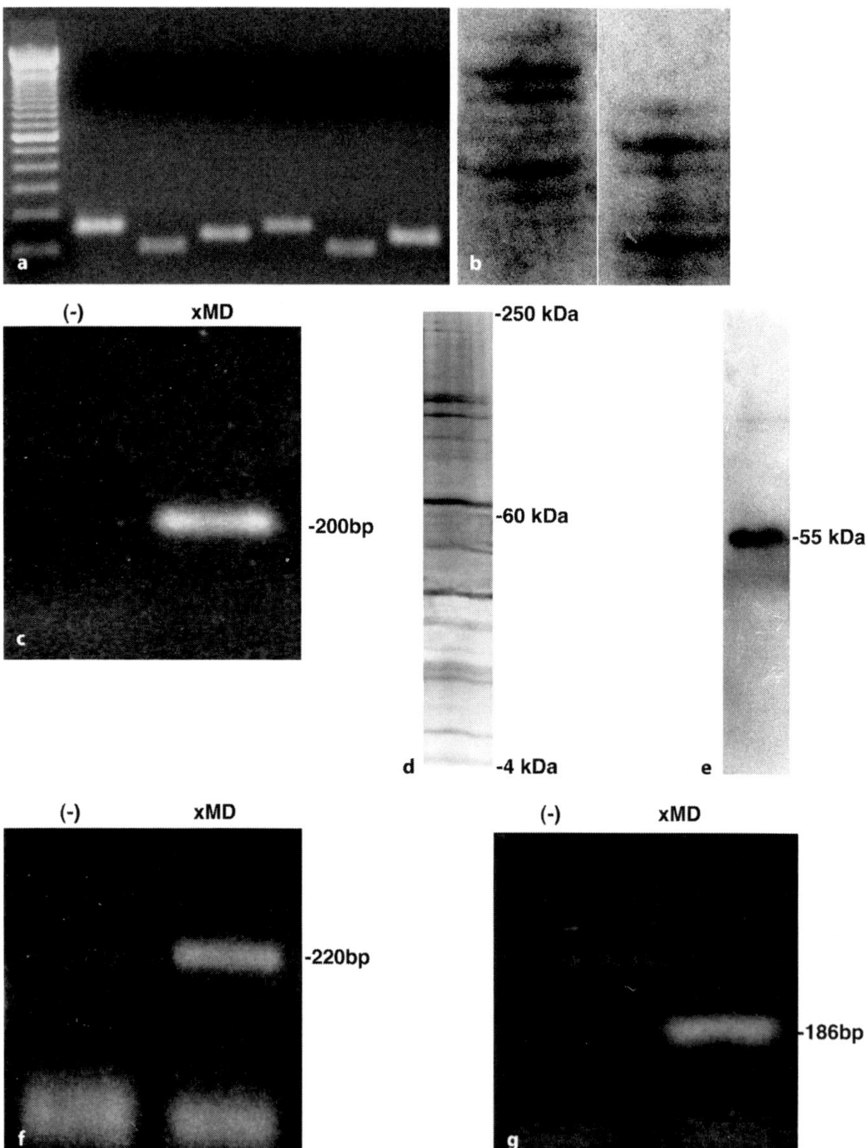

Fig. 6.5 a–g

Fig. 6.5 C. After xMD, protein lysates were evaluated by one-dimensional polyacrylamide gel electrophoresis (Fig. 6.5 D, smooth muscle actin-targeted cells) and by immunoblot for α-tubulin (Fig. 6.5 E, cytokeratin-targeted cells). Neither quantity nor quality was adversely affected by xMD. Gene-specific RT-PCR of mRNA was performed for β-actin from mRNA recovered from human prostate epithelium using cytokeratin-targeted as shown in Fig. 6.5 F, and for hypoxanthine guanine phosphoribosyl transferase (HPRT) from smooth muscle stromal cell mRNA using actin as a targeting tool (Fig. 6.5 G).

One can envision multiple embodiments of xMD. The simplest approach is to immunostain a tissue section(s), cover the slide with transfer film, and irradiate an entire field with a wide-field laser, scanning laser, or flashlamp. Employed in conjunction with an automated immunostainer, this strategy will allow rapid recovery of a large number of target cells. For example, a histological section of cancer typically contains several hundred thousand tumor cells. xMD will permit recovery of the entire cancer cell population in as many as 40 sequential tissue section recuts (10^7 or more total cells) in less than a day. Thus, investigators can study a dissected tumor cell population using analytical methods that have previously been impractical, such as those applied to cells in culture where the amount of material is not limiting. Alternatively, investigators can efficiently dissect cancers from a study set of cases to assess the frequency of particular molecular profiles across large control and analysis groups. One can also envision xMD being employed as a component of commercial laser dissection instruments, combining rapid and precise dissection with the useful features (image archiving, phenotype-based dissection) of these systems.

6.4.1 Future Directions of Expression Microdissection

There are several areas that we are actively evaluating to extend the capabilities of xMD. Firstly, the currently available transfer films used to date are prototype versions developed in our research laboratory. Although

Fig. 6.5. DNA, protein and mRNA analysis after xMD. **a** Amplification (in duplicate) of three sequence tagged site markers using DNA derived from prostate epithelium. **b** Allelotyping of prostate epithelial-derived DNA after xMD at microsatellite marker *D8S1786*. Two separate patients were analyzed both of whom were heterozygous at this locus. **c** Amplification of DNA from PCNA positive nuclei using a primer set for glyceraldehyde 3-phosphate dehydrogenase (GAPDH). **d** Analysis of total protein from smooth muscle actin (SMA) targeted cells by one-dimensional polyacrylamide gel electrophoresis, followed by silver staining. **e** Immunoblot for α-tubulin protein from cytokeratin targeted prostate epithelium. **f** RT-PCR of β-actin mRNA from cytokeratin-targeted prostate epithelium. **g** RT-PCR of hypoxanthine guanine phosphoribosyl transferase (HPRT) from smooth muscle actin positive prostate stromal cells these relatively low-quality films are capable of efficient and precise dissection, we anticipate that commercially produced, thin films optimized for xMD will significantly improve the method, possibly allowing for more precise subcellular dissection (even at better than normal optical diffraction resolution limits). Secondly, changing the amount of dye deposited on a tissue section can be accomplished by simply adjusting the primary antibody concentration, and/or the time of the enzymatic reaction that generates stain. Since there is a threshold level that results in film activation, it should be possible to use xMD to selectively procure cells that contain different amounts of target protein, in other words, high versus low expression levels. A lower intensity irradiation would bond the most highly stained targets that would be isolated on the first transfer film. Then another film placed on the same section would be irradiated at higher light levels in order to isolate a population of cells expressing the immunostain target protein at a lower level. Thirdly, the method should be readily adapted to chromogenic in-situ hybridization (CISH), allowing procurement of specific cells based on mRNA expression. Fourthly, it may be possible to utilize general, nonprobe-dependent tissue stains for xMD, although the utility may be limited due to the indiscriminate cell staining which typically occurs. Finally, in addition to dissection of cells from tissue sections, xMD should be readily applicable to cytology specimens and cultured cell samples, offering the same benefits of rapid cell procurement and dissection precision as described above for tissue sections.

We have also considered the possibility of adapting xMD to fluorescence-based targeting of cells. In this way, the technique could be multiplexed, and perhaps used in conjunction with fluorescence resonance energy transfer (FRET). Because of low backgrounds compared to transmitted light absorptive stains, fluorescence stains appear to have very high contrast for detection. However, this is a false perception when it comes to evaluating the effectiveness for absorbing sufficient light to activate a thermoplastic polymer. Thus, to date we have not had success developing a fluorescence-based application of xMD.

Another related approach to xMD is direct-label xMD. In this case, the cells of the tissue section are targeted by a probe carrying an enzyme capable of labeling the entire cellular proteome (or transcriptome) of the cells of interest. Direct-label xMD is a virtual dissection in the sense that biomolecules from the cells of interest are specifically labeled, but remain admixed with those from nontarget cells. The labeled molecules are then "dissected" via an appropriate detection system. The direct-label approach allows the investigator to physically label the biomolecules within a cell type of interest with an identifying tag. The antibody or nucleic acid probe, used as the directing agent, provides a specific delivery mechanism for the labeling enzyme, such as lactoperoxidase or reverse transcriptase (Fig. 6.6A). By using this specific delivery system, only the cells of interest obtain the labeling enzyme, therefore, when the labeling reaction is cata-

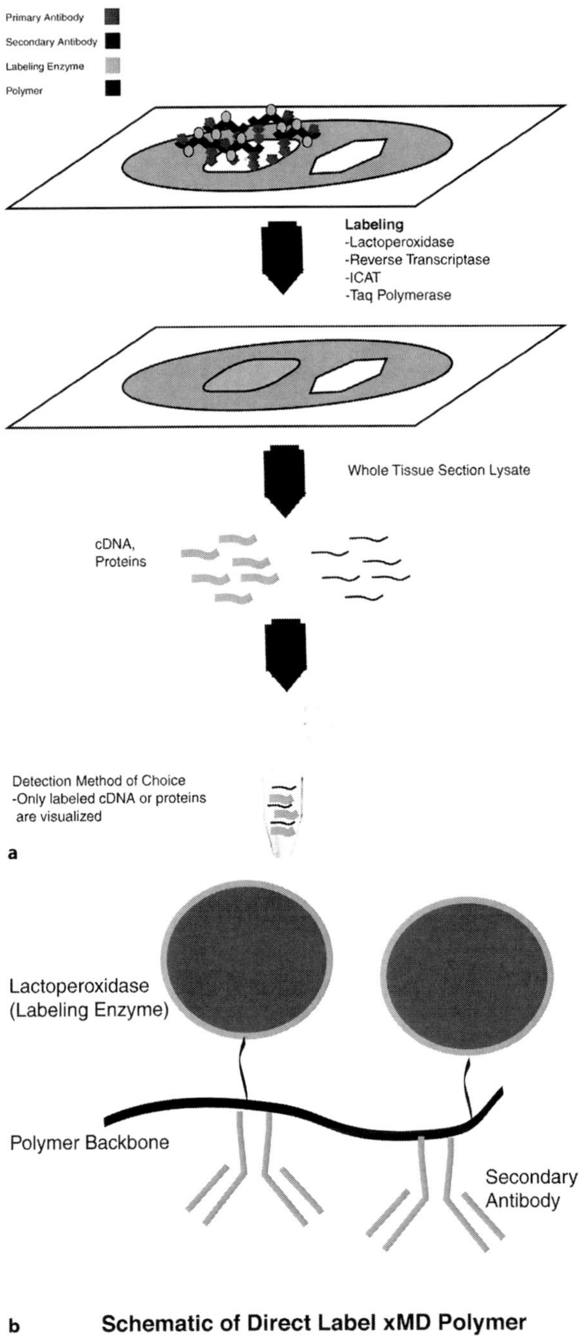

Fig. 6.6. **a** Schematic illustrating direct-label expression microdissection (xMD). **b** Simplistic illustration of a direct-label xMD polymer. The labeling enzyme is indicated by *solid circles*

lyzed, only the biomolecules within that desired cell type become labeled. The entire tissue section can then be collected and the biomolecules can be extracted. Once extracted, the labeled and unlabeled biomolecules are admixed. A detection method is then used which only detects the labeled biomolecules.

For example, a direct-label xMD polymer could be constructed which contains lactoperoxidase (an enzyme that labels tyrosine and histidine residues of proteins with I^{125}) and a detection molecule such as a secondary antibody (Fig. 6.6B). Using archived human prostate tissue, for example, the tissue section can be incubated with the direct-label xMD polymer, after a preliminary incubation with a primary antibody for prostate-specific antigen (PSA). The labeling enzyme, lactoperoxidase, is then activated by the addition of hydrogen peroxide and the labeling reaction is allowed to proceed. The labeling reaction is halted, the tissue is removed from the glass slide and placed in a protein extraction buffer. Polyacrylamide gel electrophoresis (PAGE) analysis of the sample can then be completed. Only the labeled proteome from the PSA-positive epithelial cells will be detected by autoradiography. These improvements to xMD will make the approach more robust and thus applicable to a wider audience.

In summary, expression microdissection provides a new, operator-independent method for tissue microdissection and analysis. The technique significantly increases both dissection speed and precision, and likely will facilitate studies of normal physiology, development, and disease processes.

6.5 Layered Expression Scanning

Layered expression scanning (LES) is another new technique that provides a bridge between tissue analysis and molecular profiling. The technique allows for multiple measurements of DNA, mRNA or proteins from a single biological sample (e.g., various cellular phenotypes in a tissue section, bands on a gel, individual wells of a microtiter plate, etc.; Englert et al.

Fig. 6.7. a, b Schematic illustrating the two systems of layered expression scanning (LES). In the open system (**a**), the sample is passed through the layered membrane, and each one captures a proportion of the molecules, thus generating replicates of the samples. Each membrane can be tested with a different probe or antibody. In the closed system (**b**), the membranes are coated with specific capture molecules (PCR products/clones for nucleic acids, antibodies/peptide for proteins. In the example, two messages are separated on a gel and passed through five membranes: membrane one was blocked, membrane two coated with one message, membrane three with another message, membrane four with both messages, and membrane five used as a nonspecific trap. Each membrane captures a specific molecule from the sample

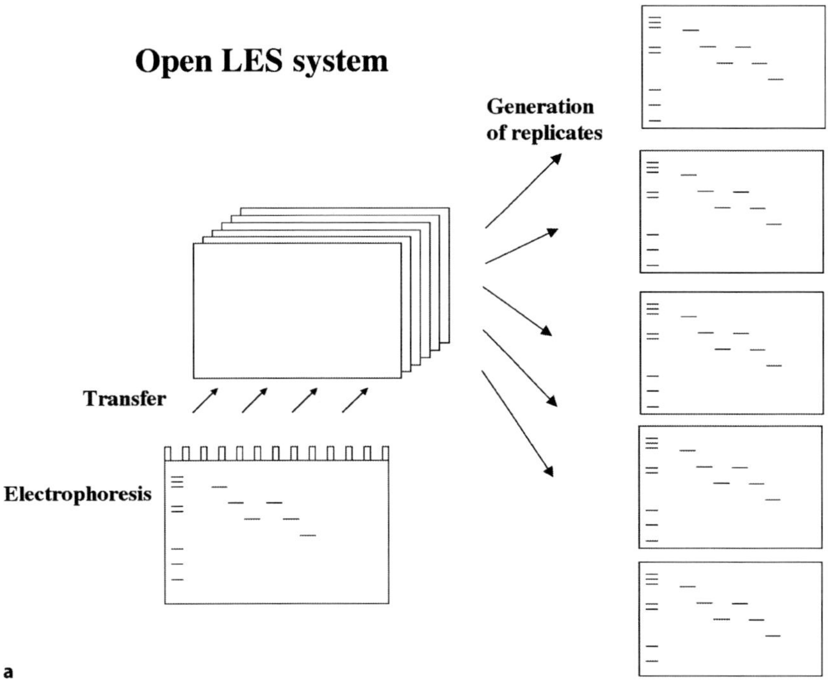

Open LES system
Generation of replicates
Transfer
Electrophoresis

a

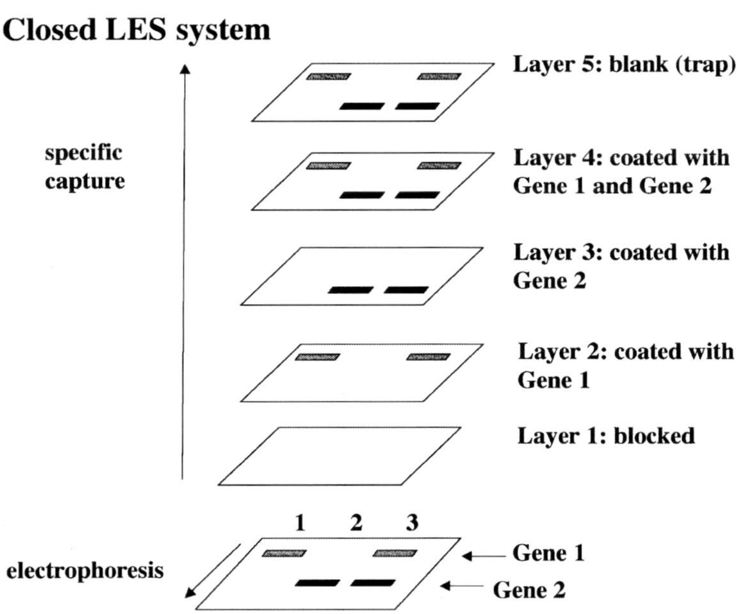

Closed LES system
Layer 5: blank (trap)
specific capture
Layer 4: coated with Gene 1 and Gene 2
Layer 3: coated with Gene 2
Layer 2: coated with Gene 1
Layer 1: blocked
1 2 3
Gene 1
electrophoresis
Gene 2

b

Fig. 6.7 a, b

2000; Tangrea et al. 2003). LES was co-developed by the National Institutes of Health and 20/20 GeneSystems, Inc. in 2000. The method is based on the transfer of biomolecules from a single sample through a series of layered membranes. As the biomolecules move through the membranes, the original two-dimensional architecture of the sample is maintained and a portion of the biomolecules is captured on each membrane. This permits the concurrent measurement of each specific sample element retained on the individual membranes. The LES method has a number of properties that increase utility; it is conceptually simple, requires no moving parts, can be used as an open or closed format, and maintains target biomolecules at a high concentration during the analysis process to produce sensitive measurements. The technique can also be applied to a variety of life science platforms, including; electrophoresis gels (one- and two-dimensional protein gels, RNA or DNA gels), tissue sections, tissue arrays, cytological samples, or multi-well plates.

In the open format LES, each membrane captures a proportion of the molecules, nonspecifically (Fig. 6.7 A). This approach allows for the generation of multiple replicates of the sample. Each membrane can then be used to analyze a different probe or antibody, increasing the output of data generated from the original sample.

In the closed system, each membrane is coated with specific capture molecules (Fig. 6.7 B). In the case of nucleic acids, gene-specific PCR products or clones can be used to coat the membranes. Since only one specific transcript or protein is captured on each membrane, the closed system offers the potential advantage that many more membranes can be used in each experiment, without obtaining a significant decrease in signals in deeper layers. The closed system also offers the advantage that the hybridization step is performed while the molecules are moving, which significantly shortens the protocol time.

6.5.1 Open Layered Expression Scanning System

Northern and immunoblots are utilized in laboratories worldwide, and offer several important features for analyzing mRNA and protein expression including accuracy, low cost, evaluation of probe specificity, and information on transcript and protein forms based on molecular size. However, standard blotting techniques are hampered by three factors: they require a significant amount of input material, are laborious, and are capable of measuring only one protein or transcript at a time. Therefore, the first goal for the development of the LES technique was to specifically facilitate multiplex analysis of RNA and protein electrophoresis gels using a layered, open system (Fig. 6.7 A; Tangrea et al. 2003). As a result, we developed a unique, highly efficient hybridization membrane that permits multiple blots to be produced from a single sample, eliminating a number of potential experimental artifacts associated with standard blotting techniques including: variance in sample preparation and solubility in loading buffer, problems associated with sample storage and/or freezing–thawing, and the typical variability seen among electrophoretic gel runs. Each blot can be probed subsequently for individual transcripts or proteins using standard detection procedures. The technique also maintains the advantages presented by standard blotting techniques, including information beyond simple expression levels such as transcript size(s) and protein processing status. The LES system makes feasible quantitative determination of the expression level of a large set of genes and proteins, such as those typically identified in array and proteomic studies.

6.5.1.1 Multiple Northern Blots

In order to assess the technical capabilities of the new LES membrane system, we evaluated the following performance parameters: hybridization characteristics, signal sensitivity, and reproducibility relative to standard blots. A series of experiments were performed using similar conditions comparing LES to standard Northern blotting experiments. Total RNA (15–30 μg) from several cell lines was electrophoresed in a 1%, denaturing agarose gel, washed and a standard Northern blot capillary transfer was performed overnight, using either a regular nitrocellulose membrane or the ten-layer LES membrane system. Measurement of total RNA content for each LES membrane showed that the maximal variability ranged between 20–30% (Fig. 6.8 A). This degree of alteration has minimal effect on subsequent probe hybridization results, and is within the normal range of experiment-to-experiment variability typically observed with standard Northern blots.

For experiments analyzing specific gene levels, the membranes were hybridized with a glyceraldehyde 3-phosphate dehydrogenase (GAPDH) probe. A specific 1.3-kb band corresponding to each of the transcripts analyzed was seen in each membrane. Quantitation of the signal showed similar results to overall RNA levels. Only a relatively small variation between the membranes is observed, ranging from 30–40%. Taken together, the total RNA and probe-specific quantitation data show that the layered array system generates multiple blots that can be successfully probed for a specific gene. Hybridizations with specific probes for HPV-18, E6/E7, PCNA, and cdc2 on three membranes using 20 μg of HeLa total RNA in a single experiment are shown in Fig. 6.8 B.

In addition to its low binding capacity characteristic, the LES membrane was also designed to provide increased hybridization efficiency. In other words, even though it binds significantly less RNA or protein, the signal intensity after probing approaches that of a traditional blot. This feature significantly expands the utility of the system, as investigators can perform

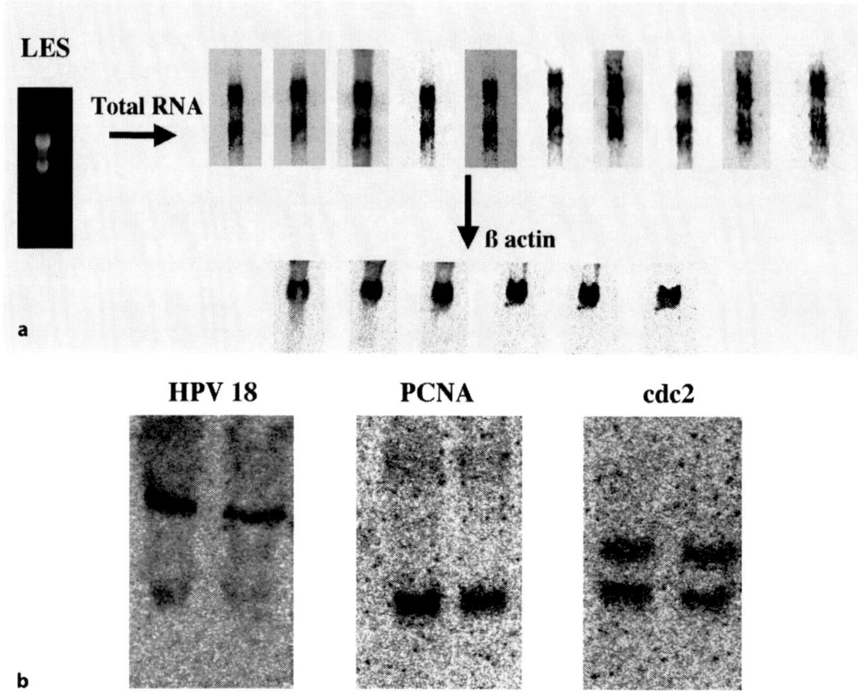

Fig. 6.8. LES Open system for RNA analysis. RNA (10 μg) was transferred through ten-layered expression scanning membranes (**a**, *above*). A typical rRNA pattern was seen in all membranes after total RNA staining. Multiple membranes were then probed for *β*-actin, all of them showing a specific signal (**a**, *below*). HPV-18, PCNA, and cdc2 expression levels using 20 μg of HeLa cell line total RNA on three layers were analyzed in a single experiment (**b**)

multiple Northern or immunoblots from a single sample preparation. To compare hybridization results between the layered array system and a standard blot, 15 μg of total osteosarcoma MG-63 cell line RNA was analyzed by both methods in parallel. After the transfer of RNA out of the gels, all membranes were probed, washed, and imaged under identical conditions. Qualitatively, the results were similar between the two blots in terms of specificity and hybridization background.

In general, we have observed that the layered array membranes produce band intensities ranging from 50–95% of traditional blots for both mRNA and proteins. For the membranes that show a 40–50% decrease in intensity, it is necessary to expose the blots to a phosphoimager, or to autoradiography film for an extended period of time to produce banding patterns that are comparable to those seen on traditional blots.

The intermembrane and intramembrane variability of the LES system has also been assessed (Tangrea et al. 2003). To evaluate intermembrane variability, we analyzed the expression ratio of the same gene (e.g., a housekeeping gene or also total RNA measurement) within the same lane between membranes. This is equivalent to repeating several Northern experiments using the same probe. The LES intermembrane variability is comparable to that observed when performing Northern repeats. The intramembrane variability has also been evaluated through the assessment of expression ratios (e.g., a housekeeping gene or also total RNA measurement), between different lanes within the LES membranes. It is important to evaluate this ratio on several membranes, since a change on one membrane could merely reflect loading variability. After thorough evaluation, it is clear that intermembrane variability within the LES system is minimal. In other words, whenever there is a fold difference in signal between LES membranes in a lane (e.g., a twofold GAPDH increase on lane 1 in membrane no. 2 compared to membrane no. 9), the same fold difference holds up for all of the other lanes within the membrane (i.e., no. 2 vs. no. 9). We recommend probing one membrane for a housekeeping gene to normalize gel loading, and then performing expression measurements of additional transcripts or proteins on the remaining layers. This is an optimal use of the system and allows data to be generated quickly and efficiently.

6.5.1.2 Multiplex-Immunoblots

Generation of multiple protein blots using the LES system was assessed by evaluating total protein staining on ten-layered membranes (Tangrea et al. 2003). Cell line protein extract was separated by polyacrylamide gel electrophoresis (PAGE) and transferred to ten-layered membranes using standard immunoblotting methods. The evaluation of total and activated levels of Raf and Erk proteins was used as a demonstration of the LES technique (Fig. 6.9). After primary incubation with the antibody of interest, all of the membranes were incubated with an anti-GAPDH antibody to confirm equal loading and transfer from the gel.

As a rule, the membranes performed similarly for the protein applications as was observed for RNA analysis (Sect. 3.1.1). The difference in intensity using total protein staining among the multiple membranes ranged from 12–22% from the average value. We analyzed the EGFR pathway in a single experiment, including total Raf protein, phosphorylated Raf, total Erk protein, and phosphorylated Erk, respectively (Fig. 6.9). The data show that the ratio of total to activated protein varies among the various cell lines, indicating that protein expression and activation status are independently regulated. Thus, it is important to measure both forms of Raf and Erk proteins to understand the overall kinetics and activation state of the pathway. Multiplex measurements such as these are likely to become increasingly important in the future as investigators move toward a systems-based understanding of biological processes.

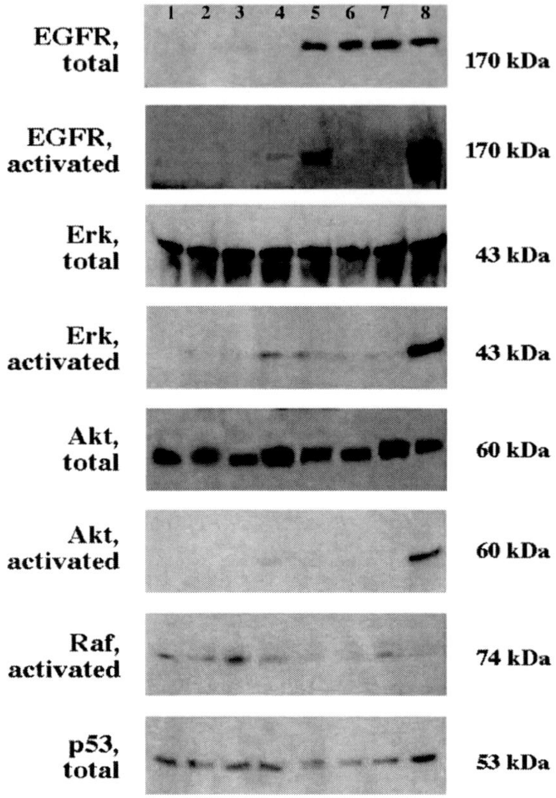

Fig. 6.9. A multilayered immunoblot showing the analysis of several members of the EGFR pathway in a single experiment of eight different cell lines HaCat (*lane 1*), Jurkat (*lane 2*), HeLa (*lane 3*), NIH3T3 (*lane 4*), SW480 (*lane 5*), MCF7 (*lane 6*), 293T (*lane 7*), S49 (*lane 8*)

6.5.2 Closed Layered Expression Scanning System

In the closed LES system, the biomolecules from the original sample are also passed through a stack of layered membranes as with the open system, but in this case the biomolecules hybridize to the membranes specifically coated with capture molecules. A schematic showing this approach is shown in Fig. 6.7B. Specific molecules (plasmid clones or PCR products for nucleic acid capture, antibodies or peptides for protein capture) are used to coat the membranes. With the use of a gel as the sample platform, isotope- or fluorescently labeled products are separated on a gel and passed through the stack of LES membranes. As the biomolecules move through the stack of membranes, only the corresponding products from the gel are captured on the appropriately coated membrane. In Fig. 6.10, an example of nucleic acid analysis using the closed LES system is shown. Two PCR

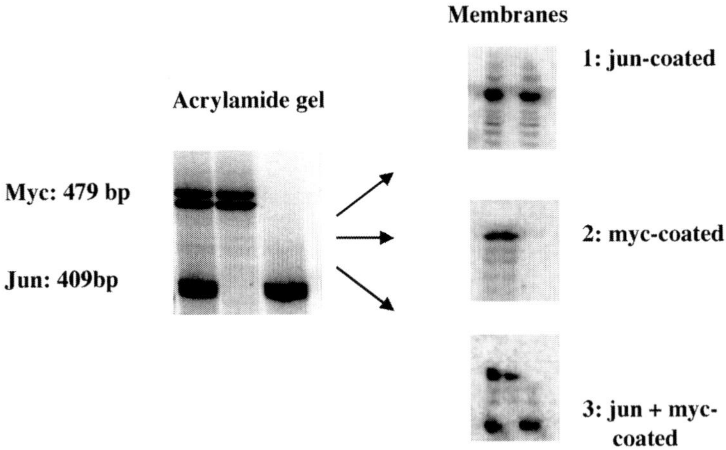

Fig. 6.10. Closed LES system. Labeled RT-PCR products for myc and jun were separated by PAGE (*left*) and passed through membranes that were coated with unlabeled products for myc, jun, and both, respectively. Specific capture of the products is shown on the corresponding membranes: jun on layer 1, myc on layer 2, and both on layer 3 (*right*)

products, one for the *myc* gene and one for the *jun* gene were generated to coat the LES membranes: one membrane was coated with *myc*, one with *jun*, and one with both (Fig. 6.10). To prevent nonspecific capture, the coating molecules were suspended in a solution containing salmon sperm and Denhardt's solution. Two radiolabeled products were separated on an acrylamide gel and then passed though the coated LES membranes. The corresponding products were specifically captured by each of the membranes (Fig. 6.10). It is important to retard the transfer of the biomolecules to allow them a longer contact time with the coated membranes, in particular with the case of nucleic acid hybridization. We have been able to achieve this by performing overnight capillary transfer. In addition, significantly reducing the amount of transfer paper allows the transfer buffer to start the movement out of the gel, but stay in contact with the membranes for longer periods of time.

Dot blots or slot blots represent ideal platforms to apply the closed LES system. The transfer can be controlled and therefore, optimal hybridizations can be achieved. Finally, antigen-antibody interactions occur on a much shorter time scale than nucleic acid hybridizations, which represents a distinct advantage for applying the closed LES system to protein analysis.

6.5.3 Other Platforms for the Layered Expression Scanning Technique

One of the most exciting LES approaches is the direct use of preserved tissue sections for protein analysis. The proteins from the tissue can be eluted and moved through the stack of membranes, retaining the original two-dimensional conformation that was present in the original sample (Fig. 6.11 A). Even though the conditions for this approach have not yet been optimized, current experiments allow for the identification of histological structures with total protein staining. For example, a whole mount, ethanol-fixed, paraffin-embedded prostate section was used to elute the proteins and pass them through the membranes. The total protein staining on one membrane shows that normal glands, hyperplastic areas, and stromal areas can be recognized (Fig. 6.11 B). After transfer, each individual LES layer can be used to assess a different protein expression level. To begin such an experiment, a hematoxylin and eosin (H & E) stained tissue section is first optically scanned and each microscopic area of interest is recorded, including normal glands, prostatic intra-neoplasia (PIN), different grades of cancer, etc. The H & E section can then be used as the reference layer, and each of the individual LES membranes treated with different antibodies. Using specifically designed software packages, each measurement on the individual layers can be determined based on the originally defined microscopic area that was annotated from the H & E image. A quantitative conversion of the staining intensity to a corresponding value can then be achieved.

The LES approach provides a variety of advantages compared to standard immunohistochemistry (IHC) analysis. For example, accurate quantitations can be achieved, since the dynamic range of intensity measurement after enhanced chemiluminescence (ECL) staining of each membrane is significantly higher than the standard on/off results obtained with in situ IHC experiments. In addition, it is common that after a few recuts deep into a tissue block, the morphology alters or the area of interest gets exhausted. Using the LES system, several measurements can be performed from a single tissue section, saving valuable clinical material and circumventing this common problem. Another advantage of the technique is that the detection of proteins using the tissue LES system is performed on proteins that are immobilized on a membrane, similar to standard Western blotting. This approach facilitates the detection of proteins compared to the antibody reaction of IHC, which occurs on a complex three-dimensional tissue section, potentially increasing the spectrum of antibodies that yield consistent signals.

Several other platforms can also be used as the sample source for LES analysis including: Southern blot gels, protein lysates, tissue microarrays, cytological samples, slot blots, hair, or microdissected cells. Some of these different platforms are illustrated in Fig. 6.12. The generation of replicates from cytological samples is especially attractive since it is impossible to

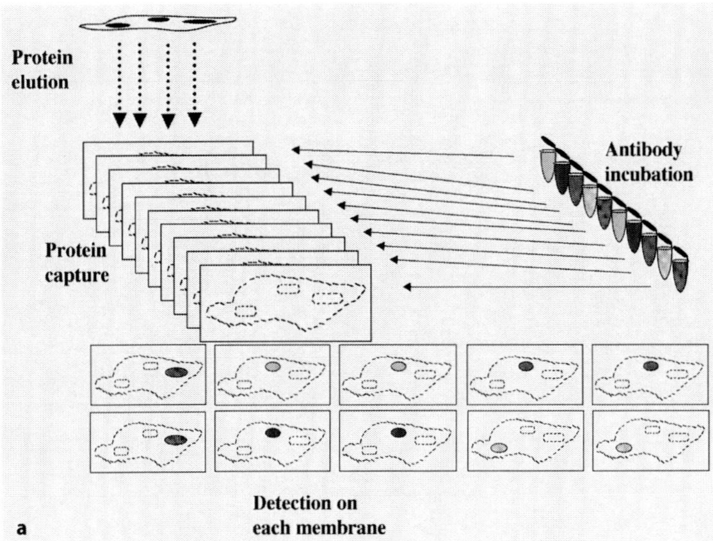

a

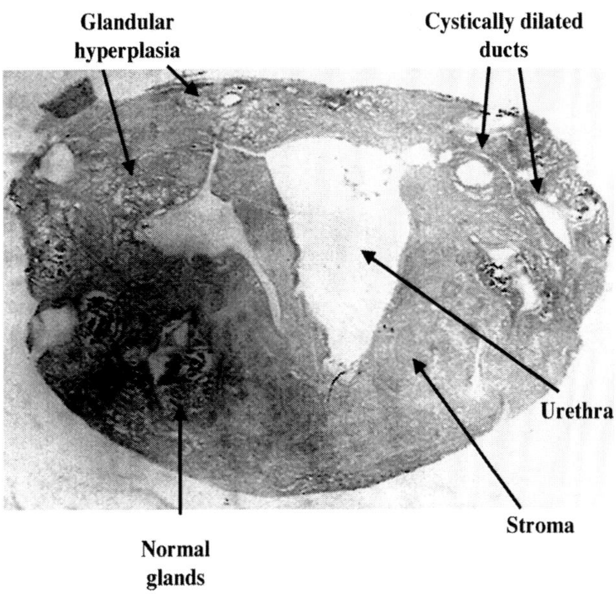

b

Fig. 6.11. Direct application of LES to tissue analysis. Proteins from the tissues are eluted, and passed through the membranes, each one capturing a proportion of the proteins. Several antibodies can then be tested in a single experiment (**a**). The proteins retain the original two-dimensional architecture of the sample. Total protein staining of a whole mount prostate section shows areas of glandular hyperplasia, dilated ducts, urethra, stroma, normal glands (**b**)

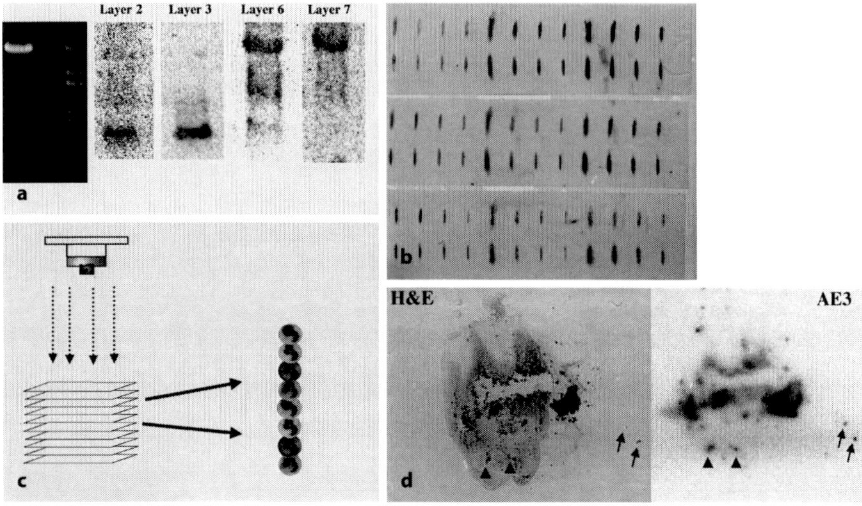

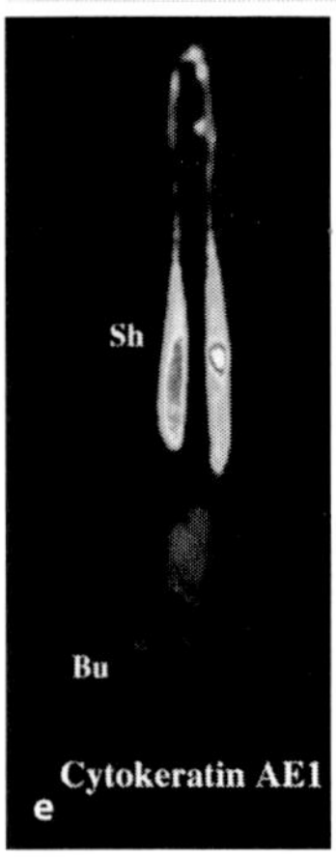

Fig. 6.12. Several biological samples can be used as platforms for LES open system analysis, including gels for Southern blots (**a**), cell lysate slot blots (**b**), laser capture microdissected cells (**c**), cytological samples (**d**), and hair follicles (**e**)

obtain recuts from these precious specimens. The only prerequisite for the use of the LES system is that the original sample must allow for the movement of biomolecules away from the sample and into the stack of LES membranes. In most cases, standard molecular biology methods can be applied to the LES system, with minimum additional effort by the researcher.

6.5.4 Future Directions

Despite the exciting prospect of applying the LES system to tissues or blotting methods, there are still some limitations to this first generation method that need to be considered. The primary issue is sensitivity. As a rule of thumb, transcripts or proteins that produce a clear signal on a traditional blot will work well in the layered array system. However, in cases where the biomolecule of interest is difficult to detect on a standard blot, it is likely that it will be similarly difficult (or impossible) to measure using the layered membranes. This limitation is inherent to the difficulty in accurately measuring low abundance mRNAs or proteins and is not unique to the LES system per se. Nonetheless, we are actively working to optimize performance of the membranes specifically for use with low abundant molecules.

For a subset of RNA and protein preparations, we have observed a less regular distribution of biomolecule content among the membranes, with a shift towards the first few layers. In these instances, the hybridization signal on membranes 1-3 often reaches 100–120% of that of standard blots, while membranes 8–10 can show levels equal to 50–60% of the expected amount. It is important to note that quantitation of transcript levels is not compromised by this effect and all ten blots produce useful data. The decrease in signal intensity for layers 8–10 can generally be overcome by using slightly more RNA or protein, for example, 25–30 μg rather than the usual 10–20 μg. Moreover, as a practical matter, it is generally recommended to probe membranes 1–6 for biomolecules of moderate (or unknown) abundance levels, and to probe membranes 7–10 for transcripts or proteins of higher abundance since these signals will be strong enough to measure even when the sample content drops to 50% of that seen on a traditional blot.

To address these issues and to improve the overall method, the addition of an amplification step to LES for nucleic acid analysis, and the antibody release LES (ARLES) approach are currently being developed in our laboratory. An amplification step is necessary to consistently apply the LES system to nucleic acid analysis directly from tissues. The total amount of RNA contained in one tissue section is well below the sensitivity of standard detection techniques. We are currently working on the application of *in situ* global transcriptome amplification protocols. This amplification step would allow for the generation of enough template to subsequently be eluted from the tissue and allow the biomolecules to be captured on the LES membranes. Even though the size information of the transcripts is lost, the retention of the two-dimensional spatial relationship would allow for the correlation of signals to each of the defined microscopic areas of interest.

The application of an amplification step could also be helpful for gel analysis. After separating the products on a gel, an "in gel" PCR could be applied, retaining the global size separation. After the PCR reaction was completed, the enriched gel would be passed through the LES membranes using the standard transfer method. This would allow for the analysis of microdissected samples on Northern blots, which until this point was difficult, if not impossible to undertake, while maintaining the size information of the transcripts.

Antibody release LES (ARLES) is another novel approach that utilizes a closed LES system for multiple gene analysis. In this case, the capture mol-

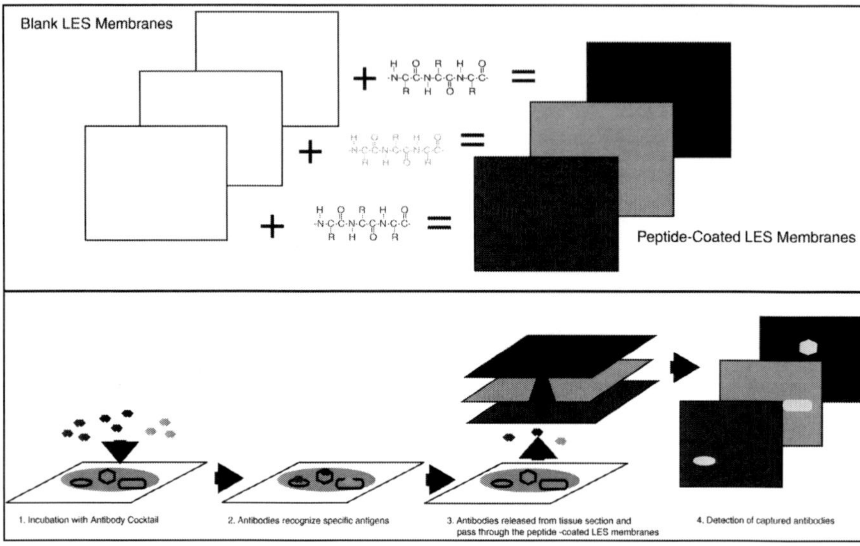

Fig. 6.13. Antibody release LES. *Above* Blank or uncoated membranes are coated with three different peptides. *Below* The steps of an antibody release LES experiment are described. The *inverted "y" of different shades of gray* represent the three different antibodies used in this example

ecules are individual peptides, with which the LES membranes are coated. Multiple antibodies are incubated on one sample (namely a tissue section) and allowed to bind to their subsequent antigens. After a series of nonspecific stringent washes, a release step is applied to the sample. The released antibodies now move out of the sample and into the stack of LES membranes. Each antibody is captured by its corresponding peptide-coated membrane (Fig. 6.13). As with standard LES, several sample platforms can be analyzed using this antibody release method, including one- and two-dimensional protein gels, 384-well plate protein lysates, tissues, etc. Moreover, the design of a polymer conjugated to a specific cDNA probe and a detection antibody would allow the method to be applied to the analysis of nucleic acids (i.e., Northern, Southern, etc.) through the mechanism of *in situ* hybridization.

6.6 Conclusion

In summary, there is an important need for novel techniques that provide a means for more specific tissue microdissection and that facilitate follow-up analysis of data from array-based studies. In this chapter we have presented two such new approaches, expression microdissection and layered expression scanning. Other such methods are also currently under development, each with particular strengths and weaknesses, for example; tissue microarrays (Kononen et al. 1998; Bubendorf et al. 1999; Dobrin and Stephan 2003), protein microarrays (Lueking et al. 1999), molecular scanners (Binz et al. 1999), and advances in mass spectrometry (Berndt et al. 1999; Yates 2000). The xMD method provides a highly precise means by which one can isolate specific cells of interest in an operator-independent manner. This new technique should have relevance to many different tissue applications. Likewise, the LES approach represents another novel methodology that also has immediate utility to researchers. The generation of gel replicates represents the first practical application of this new technology that will become valuable for the analysis of a variety of life science platforms.

References

Alizadeh AA, Eisen MB, Davis RE, Ma C, Lossos IS, Rosenwald A, Boldrick JC, Sabet H, Tran T, Yu X, Powell JI, Yang L, Marti GE, Moore T, Hudson Jr. J, Lu I, Lewis DB, Tibshirani R, Sherlock G, Chan WC, Greiner TC, Weisenburger DD, Armitage JO, Warnke R, Staudt LM et al. (2000) Distinct types of diffuse large B-cell lymphoma identified by gene expression profiling. Nature 403:503–511

Anderson NL, Anderson NG (1998) Proteome and proteomics: new technologies, new concepts, and new words. Electrophoresis 19:1853–1861

Becker I, Becker K, Rohrl MH, Hofler H (1997) Laser-assisted preparation of single cells from stained histological slides for gene analysis. Histochem Cell Biol 108:447–451

Berndt P, Hobohm U, Langen H (1999) Reliable automatic protein identification from matrix-assisted laser desorption/ionization mass spectrometric peptide fingerprints. Electrophoresis 20:3521–3526

Bersen MR, Dijkman HBPM, de Vries E, Figdor CG, Ruiter DJ, Adema GJ, van Muijen GNP (1998) Identification of multiple mRNA and DNA sequences from small tissue samples isolated by laser-assisted microdissection. Lab Invest 78(10):1267–1273

Binz PA, Muller M, Walther D, Bienvenut WV, Gras R, Hoogland C, Bouchet G, Gasteiger E, Fabbretti R, Gay S, Palagi P, Wilkins MR, Rouge V, Tonella L, Paesano S, Rossellat G, Karmime A, Bairoch A, Sanchez JC, Appel RD, Hochstrasser DF (1999) A molecular scanner to automate proteomic research and to display proteome images. Anal Chem 71:4981–4988

Bohm M, Wieland I, Schutze K, Rubben H (1997) Non-contact laser microdissection of membrane-mounted native tissue. Am J Pathol 151:63–67

Bonner RF, Emmert-Buck MR, Cole KA, Pohida T, Chuaqui RF, Goldstein SR, Liotta LA (1997) Laser capture microdissection: molecular analysis of tissue. Science 278:1481–1483

Burbach GJ, Dehn D, Del Turco D, Deller T (2003) Quantification of layer-specific gene expression in the hippocampus: effective use of laser microdissection in combination with quantitative RT-PCR. J Neurosci Methods 131:83–91

Burgemeister R, Gangnus R, Haar B, Schutze K, Sauer U (2003) High quality RNA retrieved from samples obtained by using LMPC (laser microdissection and pressure catapulting) technology. Pathol Res Pract 199:431–436

Bubendorf L, Kononen J, Koivisto P, Schraml P, Moch H, Gasser TC, Willi N, Mihatsch MJ, Sauter G, Kalleioniemi OP (1999) Survey of gene amplifications during prostate cancer progression by high-throughout fluorescence in situ hybridization on tissue microarrays. Cancer Res 59:803–806

Celis J, Kruhoffer M, Gromova I, Frederiksen C, Ostergaard M, Thykjaer T, Gromov P, Yu J, Palsdottir H, Magnusson N, Orntoft TF (2000) Gene expression profiling: monitoring transcription and translation products using DNA microarrays and proteomics. FEBS Lett 480:2–16

Chee M, Yang R, Hubbell E, Berno A, Huang XC, Stern D, Winkler J, Lockhart DJ, Morris MS, Fodor SP (1996) Accessing genetic information with high-density DNA arrays. Science 274:610–614

Chuaqui RF, Zhuang Z, Emmert-Buck MR, Liotta LA, Merino MJ (1997a) Analysis of loss of heterozygosity on chromosome 11q13 in atypical ductal hyperplasia and in situ carcinoma of the breast. Am J Pathol 150:297–303

Chuaqui RF, Englert CR, Strup SE, Vocke CD, Zhuang Z, Duray PH, Bostwick DG, Linehan WM, Liotta LA, Emmert-Buck MR (1997b) Identification of a novel transcript up-regulated in a clinically aggressive prostate carcinoma. Urology 50:302–307

DeRisi J, Penland L, Brown PO, Bittner ML, Meltzer PS, Ray M, Chen Y, Su YA, Trent JM (1996) Use of a cDNA microarray to analyse gene expression patterns in human cancer. Nat Genet 14:457–460

Dobrin SE, Stephan DA (2003) Integrating microarrays into disease-gene identification strategies. Expert Rev Mol Diagn 3:375–385

Emmert-Buck MR, Roth MJ, Zhuang Z, Campo E, Rozhin J, Sloane BF, Liotta LA, Stetler-Stevenson WG (1994) Increased gelatinase A (MMP-2) and cathepsin B activity in invasive tumor regions of human colon cancer samples. Am J Pathol 145:1285–1290

Emmert-Buck MR, Vocke CD, Pozzatti RO, Duray PH, Jennings SB, Florence CD, Zhuang Z, Bostwick DG, Liotta LA, Linehan WM. (1995) Allelic loss on chromosome 8p12-21 in microdissected prostatic intraepithelial neoplasia. Cancer Res 55:2959–2962

Emmert-Buck MR, Bonner RF, Smith PD, Chuaqui RF, Goldstein SR, Zhuang Z, Weiss RA, Liotta LA (1996) Laser capture microdissection. Science 274:998–1001

Emmert-Buck MR, Gillespie JW, Paweletz CP, Ornstein DK, Basrur V, Appella E, Wang QH, Huang J, Hu N, Taylor P, Petricoin EF 3rd (2000) A strategic approach for proteomic analysis of human tumors. Mol Carcin 27:1–8

Englert CR, Baibakov GV, Emmert-Buck MR (2000) Layered expression scanning: rapid molecular profiling of tumor samples. Cancer Res. 60:1526–1530

Fearon ER, Hamilton SR, Vogelstein B (1987) Clonal analysis of human colorectal tumors. Science 238:193–197

Fink L, Seeger W, Ermert L, Hanze J, Stahl U, Grimminger F, Kummer W, Bohle R (1998) Real-time quantitative RT-PCR after laser-assisted cell picking. Nat Med 4(11):1329–1333

Fuji H, Zhou W, Gabrielson E (1996) Detection of frequent allelic loss of 6q23-6q25.2 in microdissected human breast tissues. Genes Chromosomes Cancer 16:35–39

Going J, Lamb RF (1996) Practical histological microdissection for PCR analysis. J Pathol 179:121–124

Golub TR, Slonim DK, Tamayo P, Huard C, Gaasenbeek M, Mesirov JP, Coller H, Loh ML, Downing JR, Caligiuri MA, Bloomfield CD, Lander ES (1999) Molecular classification of cancer: class discovery and class prediction by gene expression monitoring. Science 286:531–537

Hadano S, Watanabe M, Yokoi H, Kogi M, Kondo I, Tsuchiya H, Kanazawa I, Wakasa K, Ikeda JE (1991) Laser microdissection and single unique primer PCR allow generation of regional chromosome DNA clones from a single human chromosome. Genomics 11:364–3673

Hedenfalk I, Duggan D, Chen Y, Radmacher M, Bittner M, Simon R, Meltzer P, Gusterson B, Esteller M, Kallioniemi OP, Wilfond B, Borg A, Trent J (2001) Gene-expression profiles in hereditary breast cancer. N Engl J Med 344:539–548

Klose J (1999) Genotypes and phenotypes. Electrophoresis 20:643–652

Kononen J, Bubendorf L, Kallioniemi A, Barlund M, Schraml P, Leighton S, Torhorst J, Mihatsch MJ, Sauter G, Kallioniemi OP (1998) Tissue microarrays for high-throughput molecular profiling of tumor specimens. Nat Med 4:844–847

Koreth J, Bethwaite PB, McGee JO (1995) Mutation at chromosome 11q23 in human non-familial breast cancer: a microdissection microsatellite analysis. J Pathol 175:11–18

Krizman DB, Chuaqui RF, Meltzer PS, Trent JM, Duray PH, Linehan WM, Liotta LA, Emmert-Buck MR (1996) Construction of a representative cDNA library from prostatic intraepithelial neoplasia. Cancer Res 56:5380–5383

Lee JY, Dong SM, Kim SY, Yoo NJ, Lee SH, Park WS (1998) A simple, precise, and economical microdissection technique for analysis of genomic DNA from archival tissue sections. Virchows Arch 433:305–309

Lockhart DJ, Dong H, Byrne MC, Follettie MT, Gallo MV, Chee MS, Mittmann M, Wang C, Kobayashi M, Horton H, Brown EL (1996) Expression monitoring by hybridization to high-density oligonucleotide arrays. Nat Biotechnol 14:1675–1680

Lueking AM, Horn H, Eickhoff K, Bussow H, Lehrach H, Walter G (1999) Protein microarrays for gene expression and antibody screening. Anal Biochem 270:103–111

Moskaluk CA, Kern SE (1997) Microdissection and polymerase chain reaction amplification of genomic DNA from histological tissue sections. Am J Pathol 150:1547–1552

Obiakor H, Sehgal D, Dasso JF, Bonner RF, Malekafzali A, Mage RG (2002) A comparison of hydraulic and laser capture microdissection methods for collection of single B cells, PCR, and sequencing of antibody VDJ. Anal Biochem 306:55–62

Perou C, Sorlie T, Eisen MB, van de Rijn M, Jeffrey SS, Rees CA, Pollack JR, Ross DT, Johnsen H, Akslen LA, Fluge O, Pergamenschikov A, Williams C, Zhu SX, Lonning PE, Borreson-Dale AL, Brown PO, Botstein D (2000) Molecular portraits of human breast tumours. Nature 406:747–752

Quezado MM, Moskaluk CA, Bryant B, Mills SE, Merino MJ (1999) Incidence of loss of heterozygosity at p53 and BRCA1 loci in serous surface carcinoma of the ovary. Hum Pathol 30:203–207

Radford D, Fair K, Thompson AM, Ritter JH, Holt M, Steinbrueck T, Wallace M, Wells Jr SA, Donis-Keller HR (1993) Allelic loss on a chromosome 17 in ductal carcinoma in situ of the breast. Cancer Res 53:2947–9

Rubin M (2002) Understanding disease cell by cell. Science 296:1329–1330

Schena M, Chalon D, Davis RW, Brown P (1995) Quantitative monitoring of gene expression patterns with a complementary DNA microarray. Science 270:467–469

Schena M, Heller RA, Theriault TP, Konrad K, Lachenmeier E, Davis RW (1998) Microarrays: biotechnology's discovery platform for functional genomics. Trends Biotechnol 16:301–306

Schutze K, Lahr G (1998) Identification of expressed genes by laser-mediated manipulation of single cells. Nat Biotech 16:737–740

Schutze K, Posl H, Lahr G (1998) Laser micromanipulation systems as universal tools in cellular and molecular biology and in medicine. Cell Mol Biol 44(5):735–746

Shegal D, Schiaffella E, Anderson AO, Mage RG (1998) Analysis of single B cells by polymerase chain reaction reveal rearranged VH with germline sequences in spleens of immunized adult rabbits: Implications for B cell repertoire maintenance and renewal. J Immunol 161:5347–5356

Shibata D, Hawes D, Li ZH, Hernandez AM, Spruck CH, Nichols PW (1992) Specific genetic analysis of microscopic tissue after selective ultraviolet radiation fractionation and the polymerase chain reaction. Am J Pathol 141:539–543

Speiser P, Gharehbaghi-Scnell E, Schneeberger C, Eder S, Zellinger R (1996) Microdissection as a means to verify allelic imbalance in tumor biology samples. Anticancer Res 16:461–464

Stenn KS, Paus R (2001) Controls of hair follicle cycling. Physiol Rev 81:449–494

Strausberg RL, Buetow KH, Emmert-Buck MR, Klausner RD (2000) The cancer genome anatomy project: building an annotated gene index. Trends Genet 16:103–106

Tangrea MA, Flaig MJ, Ramesh A, Best CJM, Baibakov GV, Mitchell CD, Hewitt SM, Hartmann DP, Knezevic V, Emmert-Buck MR, Chuaqui RF (2003) Layered expression scanning: multiplex analysis of RNA and protein gels. BioTechniques (in press)

Thiberville L, Payne P, Vielkinds J, LeRiche J, Horsman D, Nouvet G, Palcic B, Lam S (1995) Evidence of cumulative gene losses with progression of premalignant epithelium to carcinoma of the bronchus. Cancer Res 55:5133–5139

Wistuba II, Tang M, Maitra A, Alvarez H, Troncoso P, Pimentel F, Gazdar AF (2001) Genome-wide allelotyping analysis reveals multiple sites of allelic loss in gallbladder carcinoma. Cancer Res 61:3795–3800

Yates III, JR (2000) Mass spectrometry: From genomics to proteomics. Trends Genet 16:5–8

Zhang L, Zhou W, Velculescu VE, Kern SE, Hruban RH, Hamilton SR, Vogelstein B, Kinzler KW (1997) Gene expression profiles in normal and cancer cells. Science 276:1268–1272

Zhou M, Shah R, Shen R, Rubin MA (2003) Basal cell cocktail (34betaE12+p63) improves the detection of prostate basal cells. Am J Surg Pathol 3:365–71

Zhuang Z, Bertheau P, Emmert-Buck MR, Liotta LA, Gnarra J, Linehan WM, Lubensky IA (1995a) A microdissection technique for archival DNA analysis of specific cell populations in lesions < 1 mm in size. Am J Pathol 146:620–625

Zhuang Z, Merino MJ, Chuaqui R, Liotta LA, Emmert-Buck MR (1995b) Identical allelic loss on chromosome 11q13 in microdissected in situ and invasive human breast cancer. Cancer Res 55:467–471

7 Standard Diagnostic Techniques

Sergio González

7.1 Introduction

The conventional pathological analysis of tissue samples or biopsies enables the diagnosis of a given lesion according to current nomenclature and concepts, but new and evolving techniques are more rapidly and frequently being applied. Light microscopy, immunohistochemistry and in situ hybridization are routine diagnostic techniques in the pathology laboratory and currently constitute the standard diagnostic approach to tissue analysis. The application of new procedures from the basic research area has opened a wide and novel field of potentially diagnostic techniques. New information on the correlation between standard methods and new techniques is published every day at a vertiginous rate. The accumulated data give new information about the biology of human diseases, can be contrasted with the conventional morphological studies. Thus, the conventional diagnostic approach has become the gold standard to interpret the new information and to validate these new techniques. Pathologists are trying to keep pace and molecular biology and molecular pathology courses and seminars are attended more and more frequently by general pathologists. New generations of pathologists have formal training in molecular pathology techniques. The conventional pathology laboratory has incorporated biochemists and biologists to support the new subspecialty, i.e., molecular pathology, and to satisfy the ever-increasing needs in this field. Clinical research and routine diagnosis have obtained great benefits from these changes. In this sense, Juan Rosai's statement points in the right direction: "the creation of teams composed of basic scientists with sufficient understanding of pathology, (surgical) pathologists with sufficient understanding of basic science, and hybrids with a good formation in both areas, working together towards a common goal" (Rosai 1997).

Because neoplasias comprise the most frequent and important group of lesions investigated in the pathology laboratory, most of the new applications of modern techniques are in the field of tumor diagnosis. Enormous

Principles and Practice
Emmert-Buck et al. (Eds.):
Dissecting the Molecular Anatomy of Tissue

advances in the comprehension of tumor pathology have been made in the last few years.

7.2 Diagnostic Histopathology

Histopathological analysis is conventionally carried out by means of the light microscope and 80–90% of the lesions can be reliably diagnosed using this method. It allows the identification of neoplastic and non-neoplastic conditions and the classification of tumor types, histological grading and tumor stage. Certainly, this is a low-cost and very effective technique, yielding valuable information. Moreover, it is rapid, simple, and reliable and, although the assessment of histopathological samples is subjective and reproducibility is still a matter of concern (Trotter and Bruecks 2003), the diagnostic error rates are very low. On the other hand, current therapeutic modalities require more information about the tissular, cellular and molecular status in a given lesion, so treatment can be adjusted to individual patients. The information obtained from histopathological diagnoses does not always cover all these aspects. Tumor cells in the different phases of tumor progression express a changing constellation of cellular antigens and show a functional dynamic profile not visible under the microscope, except for changes in the differentiation grade and the appearance of metastasis. Currently, this is relevant information for therapeutics. Morphological features alone cannot explain different behavior and therapy sensitivity for certain tumors. For instance, viral lytic infections produce characteristic microscopic features such as nuclear inclusions, but latent viral infections with a low number of viral copies and/or nonlytic infections in nonpermissive cells show no morphological abnormalities identifiable in tissues, the exception being HPV infection with so-called koilocytosis changes.

7.3 Immunohistochemistry

The introduction of immunohistochemistry now offers the possibility of assessing the so-called tumor markers, in order to identify cellular differentiation of tumors and especially hormone-therapy targets and the new monoclonal antibody therapy target antigens (Fig. 7.1), as well as a phenotypic profile characteristic and diagnostic importance for a given tumor. Immunohistochemistry has become a routine diagnostic technique and has replaced conventional histochemical or special stains to identify different components of tissue sections, in particular molecules (Dodson 2000; Table 7.1). In addition, immunohistochemistry is a sensitive and specific technique that can be applied to virtually any immunogenic molecule. A

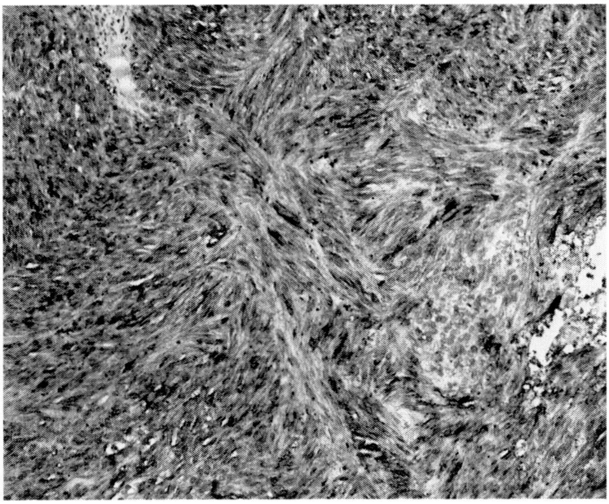

Fig. 7.1. Gastrointestinal stromal tumor showing intense positive reaction for CD117 (c-kit); S-ABC technique with hematoxylin counterstaining. Original magnification ×200

Table 7.1. Immunohistochemical detection systems

Labeled	Direct conjugate-labeled antibody method
	Indirect (sandwich) method
	Avidin-biotin conjugate method (ABC)
Unlabeled	Enzyme bridge technique
	Peroxidase-antiperoxidase method (PAP)
	Biotin-avidin method
	Biotin-streptavidin method
	Alkaline phosphatase anti-alkaline phosphatase method (APAAP)

standard protocol for p53 immunohistochemistry staining is shown in Chapter 11 (Sect. 11.5.3.3.2).

The immunophenotypic profile is the spectrum of antigens that are constant and characteristic of a given tumor. Thus, the presence of a series of antigens identified by monoclonal antibodies in biopsy samples allows the identification of a defined cellular type and tumor differentiation. This situation is exemplified by the anaplastic large cell lymphoma typically showing CD30 (Ki-1) antigen (Fig. 7.2), and also other major lymphoma groups of T-cell (CD3+), B-cell (CD20+), NK-cell (CD56+), etc.

Among the many advantages of immunohistochemistry, it is worth mentioning that it can be applied to tissues with standard fixation in formalin and paraffin embedding. Thus, it can be widely used in routine diagnosis and also applied to archival tissues. Although fixation can alter the avail-

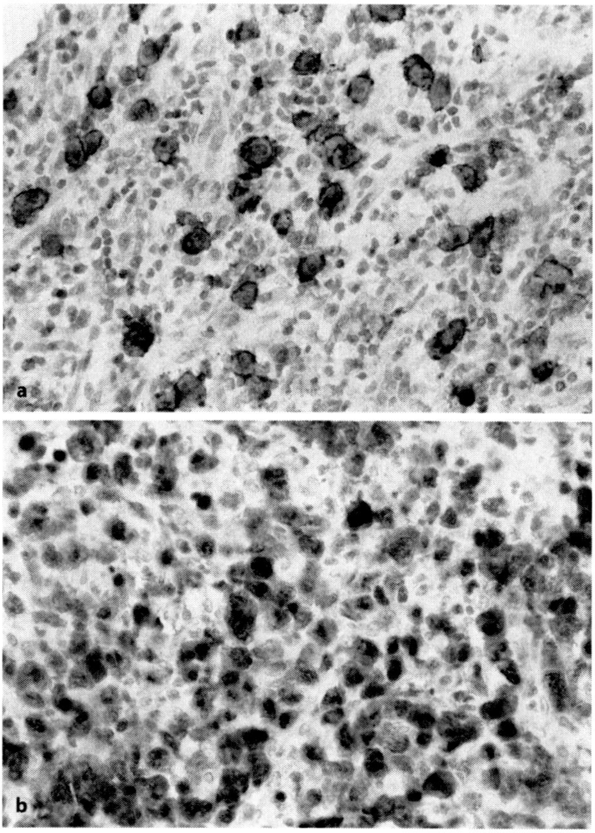

Fig. 7.2 a, b. Anaplastic large cell lymphoma of lymph nodes. **a** Immunohistochemical staining for CD30 (Ki-1) showing intense membranous signal with paranuclear dots. **b** ALK-1 intense staining in nuclei and cytoplasm of tumor cells. S-ABC technique with hematoxylin counterstaining. Original magnification ×400

ability of some antigens, reliable antigen unmasking procedures have been developed in the last few years (Miller et al. 2000). These retrieval techniques use methods to increase accessibility of antigens that are "masked" by formalin fixation and paraffin embedding, without damage to histological sections and cellular details. However, care should be taken because some antigens are modified by antigen retrieval and false negative results can be obtained (Miller et al. 2000). Immunohistochemistry has reached an acceptable level of automation that has eliminated the technical drawbacks of manually performed procedures and antigen unmasking problems. However, interpretation of immunohistochemical results is still a problem because variable and unexpected patterns of immunostaining, especially in tumors, do occur. Immunoreactive entrapped nonneoplastic cells, antigen diffusion out of normal cells to tumor cells, and nonspecific adsorption of antibodies may account for this problem. Automation includes application of reagents, incubation, and rinsing under controlled temperature and humidity conditions as well as strict timing. Sectioning and mounting are still manually performed. Automation allows in-house reproducibility, accurate results, and shorter overall times. Poor tissue fixation or overfixation, poor sectioning, and selection of inadequate areas for staining are still problems that occasionally occur. Protocolization of tissue handling and processing and coordination with pathologists before biopsy to optimize the tissue analysis are measures that ensure optimal results (Dabbs 2002).

One of the most valuable advantages of immunohistochemistry is the possibility of evaluation with optimal preservation of tissue morphology and cellular details, allowing precise identification of cells actually showing positive and/or negative reactions, including proportion and distribution of these cell populations.

Among the disadvantages, it can be mentioned that cross-reactions may occur, poor preservation of tissues can give false negative results, and many antigens are not as specific for certain tissues as initially thought. Great variability in protocols and processing among different laboratories exists, so protocol standardization remains limited to single laboratories.

Semiquantitative analysis is routinely used in estrogen and progesterone and c-erbB-2 evaluation in breast cancer (Fig. 7.3), as well as in androgen receptor in prostate cancer, and lymphoid subpopulations in lymphoproliferative disorders of nodal and extranodal sites. Quantitative immunohistochemistry is also used in other routine situations like estimation of the proliferative index with staining for Ki-67 (MIB-1), p53 staining, and lymphoid subpopulations for mycosis fungoides diagnosis (CD4+, CD8–, CD7–; Seidal et al. 2001; Dabbs 2002).

A selection of specific cell populations can be made for tissue microdissection and, properly guided by immunohistochemical staining, adequate tissues can thus be selected for genetic studies. The target cell populations are identified by previous immunohistochemistry and the selected cell subpopulations can be used for further molecular studies. Immunohistochemistry can aid in correlation studies of molecular analysis and histopathology studies. Molecular analyses are increasingly used as complementary techniques in the study of many human diseases, for instance, clonality analysis of lymphoid disorders. The diagnostic value of clonality can be contrasted with the immunohistochemical results and conventional morphological analysis.

A combination of tissue microarrays and immunohistochemistry enables a rapid and cost-effective screening of marker expression (Bubendorf et al. 2001). Up to 1000 tumor samples can be arranged in a single paraffin block, allowing the screening of a large number of tumors under the same procedure conditions (see also Chaps. 10 and 11).

These markers correlate with disease course and treatment susceptibility (CD20, CD117, CD41, CD25, ALK-1) and may help to identify tumor sub-

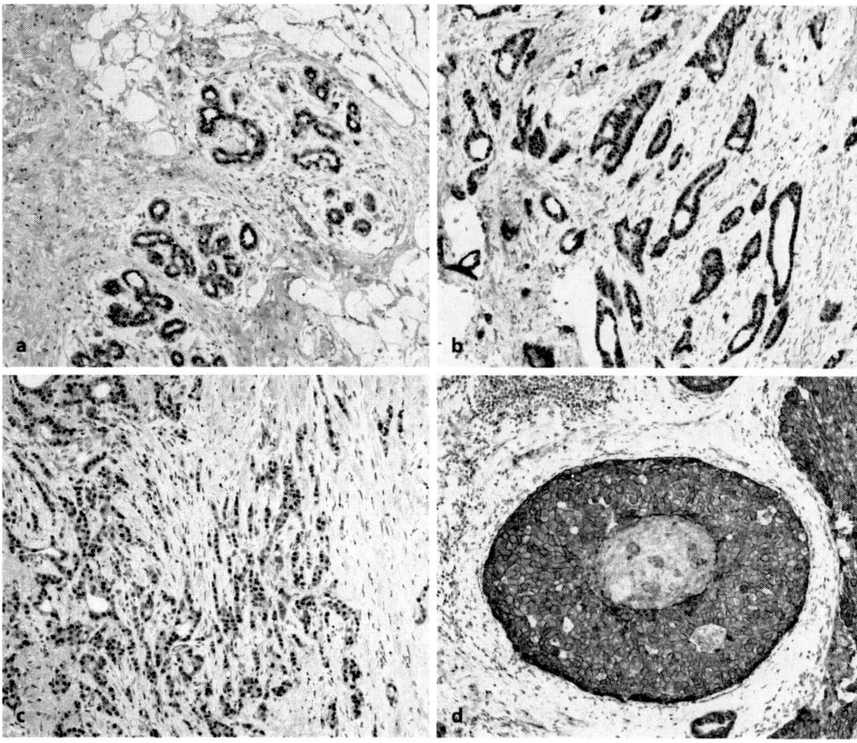

Fig. 7.3 a–d. Breast cancer. **a** Normal mammary acini showing nuclear staining for estrogen receptor; original magnification ×200. **b** Ductal infiltrating carcinoma showing nuclear staining in 100% of tumor cells (grade 3); original magnification ×200. **c** Ductal infiltrating carcinoma showing nuclear staining in 80% of tumor cells for progesterone receptor; original magnification ×200. **d** Intraductal carcinoma showing intense membranous staining for c-erbB-2; original magnification ×400. S-ABC technique with hematoxylin counterstaining

sets with particular aggressive behavior. ALK-1 or anaplastic lymphoma kinase 1 is present in about 40–70% of anaplastic large-cell lymphomas of T-cell and null-cell types. It is a chimeric protein NPM-ALK localized to the cytoplasm and nucleus of tumor cells, a product of t(2;5)(p23;q35) translocation involving the nucleophosmin (NPM) gene and a gene coding for a receptor tyrosine kinase (ALK; Morris et al. 1994). ALK immunostaining is rapid, specific, and inexpensive. Normally, it is expressed in nerve cells, but not in lymphoid cells. Translocations of alk result in overexpression of ALK and thus the immunostaining in tumor cells provides indirect evidence for abnormalities and status of the alk gene.

Immunohistochemical detection of metastatic melanoma in sentinel lymph nodes upstages 10% or more patients with occult metastases, i.e., metastases not detectable by light microscopy and immunohistochemistry using antibodies against protein S-100 and HMB-45 and/or Melan-A antigens (Perrott et al. 2003). RT-PCR to detect melanoma-associated mRNA of tyrosinase, MART-1, and gp100 have been successfully used. RT-PCR assay has shown a positive reaction in 90% of initially negative cases by conventional microscopy and immunohistochemistry. Patients with expression of two or more markers by RT-PCR have shown significantly higher risk of recurrences (Bostick et al. 1999).

7.4 In Situ Hybridization and PCR-in Situ Hybridization

In situ hybridization allows the detection of specific nucleic acid sequences, thereby permitting precise cellular location. It is used to diagnose infectious diseases and to assess gene expression in certain neoplastic disorders.

Tissues fixed in formalin and embedded in paraffin are cut and sections are mounted on silanized glass slides. Protease predigestion allows accessibility to tissue nucleic acids. Frozen sections do not need protease digestion. Probes labeled with nonisotopic compounds are suitable because they produce lower background, greater resolution, and prolonged stability of reactions. Signals are easily detected in tissues after colorimetric reactions (Fig. 7.4). Controls for ISH included hybridization of the same tissue with a probe complementary to the test probe and identical to the target DNA (same probe) and which serves as a negative control. ISH has been widely used to identify specific viral sequences in tumors, especially human papillomavirus, Epstein-Barr virus, hepatitis virus family, and HIV (Fig. 7.4). A protocol for nonisotopic in situ hybridization is shown in Chapter 11 (Sect. 11.5.2.3).

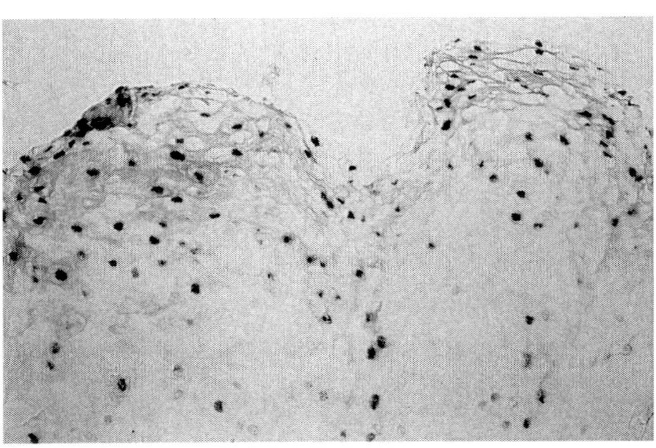

Fig. 7.4. In situ hybridization for HPV in exocervical epithelium. Intraepithelial neoplasia, grade 2, with nuclear signal in the upper layers of the atypical epithelium; APAAP technique without contrast. Original magnification ×200

ISH is less sensitive than amplification methods, but provides important histologic information. ISH cannot detect latent viral infection due to the low number of copies within infected cells (Nuovo and Cottral 1989; Nagai et al. 1987). The detection threshold of ISH is ten copies per cell (Nuovo 1997). ISH is an excellent method for the analysis of gene therapy tissues to determine the localization, distribution and expression of gene therapy vectors administered to patients.

PCR-ISH combines the sensitivity of nucleic acids amplification and the tissue localization of ISH reactions (Nuovo 1997). Limitations of this technique are related to drawbacks of PCR and ISH themselves. For instance, ISH is limited by background staining and false positive reactions, so controls are necessary for every reaction.

The main advantages of PCR-ISH are the combination of two powerful techniques such as PCR (sensitivity) and ISH (morphology preservation), information related to the localization of the reaction with pathogenic and etiologic consequences, and the easy identification of false positive results because of the possibility of localizing the signal in cells and thus discriminating true from false reactions.

Recently, RT-PCR-ISH has been applied to identify T-cell receptor beta rearrangements in cutaneous T-cell infiltrates (Nuovo et al. 2001; Magro et al. 2003). This technique allows the direct correlation of the molecular data with histological features.

7.5 Microdissection

Laser capture microdissection (see also Chaps. 10 and 11) is used to isolate defined cell populations. At present, array technologies cannot distinguish whether a lesion is an in situ carcinoma, a microinvasive carcinoma, or an invasive carcinoma with lymphatic spread, based exclusively on genetic profiles. This technology cannot per se differentiate between primary and secondary tumors (metastasis).

Any study of genetic expression must be completed with an analysis of protein expression, because we know that, after protein synthesis, modifications appear that could invalidate extrapolations from one type of study to another. Genetic microarray analysis allows the characterization of many types of genes, but it does not reveal levels of protein expression or post-translational modifications of proteins.

The adaptation of tumor cells is a known phenomenon showing dynamic transformation of the expression of proteins, giving variable protein expression profiles that can be analyzed through proteomic techniques (see Chap. 9). A combination of genetic expression analysis and the assessment of the proteomic profile, measured quantitatively and qualitatively, may lead to the identification of biomarkers for potential clinical application. Correlation of genetic studies and conventional morphology is mandatory in order to obtain adequate information of clinical relevance and regarding the biology of neoplasia.

7.6 Closing Remarks

Molecular techniques will still have to undergo a certain period of quality control (Lakhani and Ashworth 2001), and conventional techniques including light microscopy, immunohistochemistry, and in situ hybridization play an important role as standard and reference procedures. Moreover, the pathologist is the specialist in charge of diagnosing lesions and linking and translating the conceptually and scientifically sound basic research techniques with potential application in clinical diagnosis for the benefit of the patients.

References

Bostick PJ, Morton DL, Turner RR, Huyhn KT, Wang HJ, Elashoff R, Essner R, Hoon DS (1999) Prognostic significance of occult metastases detected by sentinel lymphadenectomy and reverse-transcriptase-polymerase chain reaction in early-stage melanoma patients. J Clin Oncol 17:3238–3244

Bubendorf L, Nocito A, Moch H, Suater G (2001) Tissue microarray (TMA) technology: miniaturized pathology archives for high-throughput in situ studies. J Pathol 195:72–79

Dabbs DJ (2002) Diagnostic immunohistochemistry. Churchill-Livingstone, New York

Dodson A (2002) Modern methods for diagnostic immunocytochemistry. Curr Diagn Pathol 8:113–122

Lakhani SR, Ashworth A (2001) Microarray and histopathological analysis of tumours: the future and the past? Nat Cancer 1:151–157

Magro CM, Nuovo GJ, Crowson AN (2003) The utility of the in situ detection of T-cell receptor beta rearrangements in cutaneous T-cell-dominant infiltrates. Diagn Mol Pathol 12:133–141

Miller RT, Swanson PE, Wick MR (2000) Fixation and epitope retrieval in diagnostic immunohistochemistry: a concise review with practical considerations. Appl Immunocytochem Mol Morphol 8:228–235

Morris SW, Kirstein MN, Valentine MB, Dittmer KG, Shapiro DN, Saltman DL, Look AT (1994) Fusion of a kinase gene, ALK, to a nucleolar protein gene, NPM, in non-Hodgkin's lymphoma. Science 263:1281–1284

Nagai N, Nuovo GJ, Friedman D, Crum CP (1987) Detection of papillomavirus nucleic acids in genital precancer with the in situ hybridization technique: a review. Int J Gynecol Pathol 6:366–379

Nuovo GJ (1997) PCR in situ hybridization. Protocols and applications, 3rd edn. Lippincott-Raven, New York

Nuovo GJ, Cottral S (1989) Occult infection of the uterine cervix by human papillomavirus in postmenopausal women. Am J Obstet Gynecol 160:340–344

Nuovo GJ, Morrison C, Porcu P, Caliguiri M, Suster S (2001) In situ determination of T-cell receptor beta expression patterns. J Histochem Cytochem 149:139–146

Perrott RE, Glass LF, Reintgen DS, Fenske NA (2003) Reassessing the role of lymphatic mapping and sentinel lymphadenectomy in the management of cutaneous malignant melanoma. J Am Acad Dermatol 49:567–588

Rosai J (1997) Pathology: a historical opportunity. Am J Pathol 151:3–6

Seidal T, Balaton AJ, Battifora H (2001) Interpretation and quantification of immunostains. Am J Surg Pathol 25:1204–1207

Trotter MJ, Bruecks MK (2003) Interpretation of skin biopsies by general pathologists. Diagnostic discrepancy rate measured by blinded review. Arch Pathol Lab Med 127:1489–1492

8 DNA Applications

Pablo Uribe and Ignacio I. Wistuba

8.1 Introduction

The study of normal and neoplastic genomes has been the most widely used application of microdissected material in the study of cancer pathways (Gillespie et al. 2001). Genetic changes in the multistep progression of cancer can involve amplification or gain of function mutations in dominant oncogenes, or they may involve loss of function by deletion, mutation or methylation in recessive tumor suppressor genes. According to Knudson's classical two-hit hypothesis of tumor suppressor gene function, one parental allele is lost by deletion, while the second is inactivated by mutation (Knudson 1985). Thus, if tumors are analyzed with respect to the integrity of their parental alleles at a given polymorphic locus, both alleles would be present in the constitutional DNA while one allele would be lost in the tumor (a phenomenon called "loss of heterozygosity" or LOH). Microdissection has made a remarkable difference in the application of LOH analysis to the study of cancer pathways, specially in formalin-fixed and paraffin embedded tissues (FFPET; Maitra et al. 2001). The logistics of LOH analysis are such that virtually pure populations of tumor cells or preneoplastic foci are required, since contamination by even a few unwanted cells will mean the second allele deleted in the cell population of interest will be amplified by the PCR reaction. LOH analysis has been invaluable for mapping tumor suppressor genes (TSGs), localization of putative chromosomal "hot spots" and the study of sequential genetic changes in preneoplastic lesions. The use of microdissected material from preneoplastic lesions has shown that many of the genetic alterations found in cancers actually begin in histologically normal appearing tissue (Wistuba et al. 1997b).

Besides LOH analysis, other studies that can be performed on microdissected DNA include analysis of X-chromosome inactivation to assess clonality, single strand conformation polymorphism (SSCP) and restriction fragment length polymorphism (RFLP) analyses for mutations in critical genes, comparative genomic hybridization (CGH) and the analysis of pro-

Principles and Practice
Emmert-Buck et al. (Eds.):
Dissecting the Molecular Anatomy of Tissue

moter hypermethylation (Maitra et al. 2001). All of these methods are applicable to FFPET. In some of them no more than 100–200 sectioned cells are required per PCR reaction and even fewer cells if material from cryostat sections or methanol-fixed specimens is used. CGH, which until recently was limited by large quantities of fresh/frozen DNA, has been successfully performed using 20–100 microdissected cells (Taguchi et al. 1997; Paris et al. 2003). The combination of microdissection with newer technologies such as primer extension pre-amplification and whole genome amplification will make genomic analyses possible on smaller quantities of cells, even from FFPET, thereby permitting the study of microscopic normal cells and preneoplastic lesions (Wang et al. 2001; Heinmoller et al. 2002).

DNA extracted from FFPET has been considered to be a poor substrate for molecular studies (Diaz-Cano and Brady 1997; Legrand et al. 2002). However, in this chapter, we will discuss different successful methodologies for genetic analysis of FFPET-derived DNA based on PCR amplification, the method used most frequently in the molecular analysis of this type of sample. Their applications as well as the underlying causes of methodological problems and the potential techniques to overcome them will also be reviewed.

8.2 DNA Extraction

There are several protocols available for DNA extraction after microdissection from FFPET (de Lamballerie et al. 1994; Hung et al. 1995; Coombs et al. 1999; Sato et al. 2001; Legrand et al. 2002). The most common methodology used is that based on proteinase K digestion. For proteinase K digestion to extract DNA from microdissected FFPET, important technical issues that need to be addressed are discussed in the Protocols section (Chap. 11.4.1.2).

8.3 DNA PCR Amplification

All the genomic applications for FFPETs discussed in this section are based on the ability to successfully amplify DNA sequences using PCR-based strategies. It has been demonstrated that for PCR DNA amplification, the success rate decreased proportionally with the time of formalin fixation and, consequently, with the size of the required DNA template (Legrand et al. 2002). Short alleles have shown higher efficiencies than long alleles, and uniplex reactions perform much better than multiplex amplifications. The main role in DNA degradation has to be attributed to formalin fixation time. Formalin-induced degradation starts early, after the first 3 h of fixation, even in phosphate buffered formalin (Legrand et al. 2002). However, successful amplifications have been reported for fixation times varying from 1 to 7 days, and up to 70 days (Savioz et al. 1997). There is abundant scientific literature regarding PCR amplification in this type of DNA, therefore, a few basic practical key aspects of this procedure are shown in the Protocols section (Chap. 11.5.1.3).

8.4 Microsatellite Analysis

Microsatellites are highly polymorphic short tandem repeat (dinucleotide and multinucleotide) DNA sequences. Several of these microsatellite markers have been used extensively for DNA fingerprinting and are very useful in genetic linkage analysis. Loss of heterozygosity (LOH) analysis utilizing polymorphic microsatellite markers is frequently used to identify allelic losses at specific chromosomal loci in tumors and precursor lesions (Wistuba et al. 1999, 2000, 2001, 2002). In addition to those specific genetic changes, other evidence indicates that genetic instability occurs in human cancers. This evidence includes changes in the length of microsatellite markers, frequently present in a wide variety of cancer types, and is known as microsatellite instability (MSI; Boland et al. 1998; Wistuba et al. 1998). MSI represents changes in the size of simple nucleotide repeat polymorphic microsatellite markers, resulting in altered electrophoretic mobility of one or both alleles, associated with a deficiency in the DNA mismatch repair system. The development of methodologies for PCR-based assays in

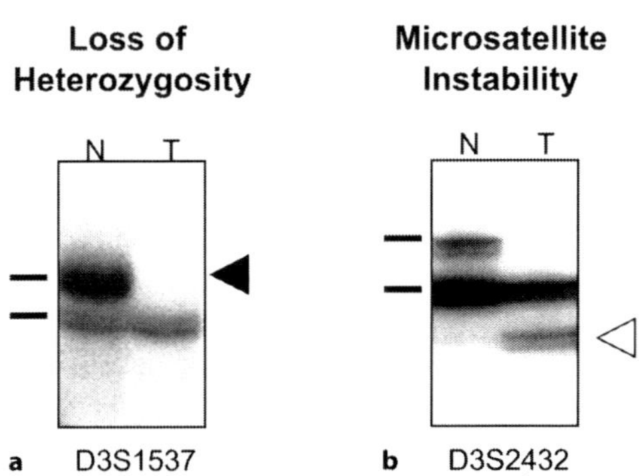

Fig. 8.1. Examples of loss of heterozygosity (LOH) and microsatellite instability (MSI) analyses of DNA extracted from microdissected FFPET obtained from lung cancer specimens. Both panels demonstrate microsatellite analysis of microdissected DNA of paired nonmalignant (*N*) and tumor (*T*) tissue samples from individual patients. **a** Tetranucleotide repeat marker D3S1537 (located in chromosome 3p22–24.2 region), demonstrating LOH of the upper allele in the tumor sample; **b** tetranucleotide repeat marker D3S2432 (3p22–24.2), demonstrating loss of one tumor allele and a mobility shift (MSI) of the remaining allele

small number of cells from FFPETs, and of precise microdissection techniques have facilitated the analyses of LOH and MSI in tissues processed using conventional histopathological techniques (Boland et al. 1998; Maitra et al. 2001). Primers amplifying specific and known microsatellites are available in the Genome Database website (http://www.gdb.org).

For LOH and MSI analyses in a particular case, constitutional microsatellite pattern is determined by examination of paired nonmalignant tissue (Maitra et al. 2001). LOH is scored by visual or image analysis detection of complete absence or reduction of the one tumor allele in heterozygous (i.e., "informative") cases (Fig. 8.1). On the other hand, MSI is detected by a shift in the mobility of one allele, irrespective of whether it is accompanied by LOH. MSI either consists of deletions (recognized by increased electrophoretic mobility), or insertions (recognized by decreased electrophoretic mobility; Fig. 8.1).

The basic PCR reaction used by us for amplification and electrophoresis analysis of polymorphic microsatellite DNA sequences in our LOH and MSI studies is summarized in the Protocols section (Chap. 11.5.1.4).

8.5 Mutation Analysis

Several strategies have been used to screen and confirm gene mutations in DNA extracted from FFPET (Frayling 2002). The most frequently used screening methods for PCR-based mutation analyses used in this type of material, SSCP (single-strand conformation polymorphism) and RFLP (restriction fragment length polymorphism) analyses, will be discussed in this chapter.

8.5.1 Single-Strand Conformation Polymorphism

This method has been designed to detect DNA sequence changes as minor as a single base difference. The conformation of single-stranded DNA, under nondenaturing conditions, is dependent mainly on intra-strand interactions. Consequently, the mobility characteristics of the single-strand DNA in polyacrylamide gel electrophoresis will be based on its nucleotide sequence. In this technique, genomic DNA PCR amplified fragments are denatured and examined under nondenaturing polyacrylamide gel conditions (Fig. 8.2). Electrophoretic mobility changes of the DNA bands have been found to be indicative of base sequence changes. Single base changes in DNA fragments up to 200 nucleotides long can be detected as mobility shifts. The characteristics of SSCP allow one to analyze specific amplified DNA sequences extracted from FFPET, and it has been applied to screen mutations in a number of genes and tumor types (Wistuba et al. 1997a; Frayling 2002).

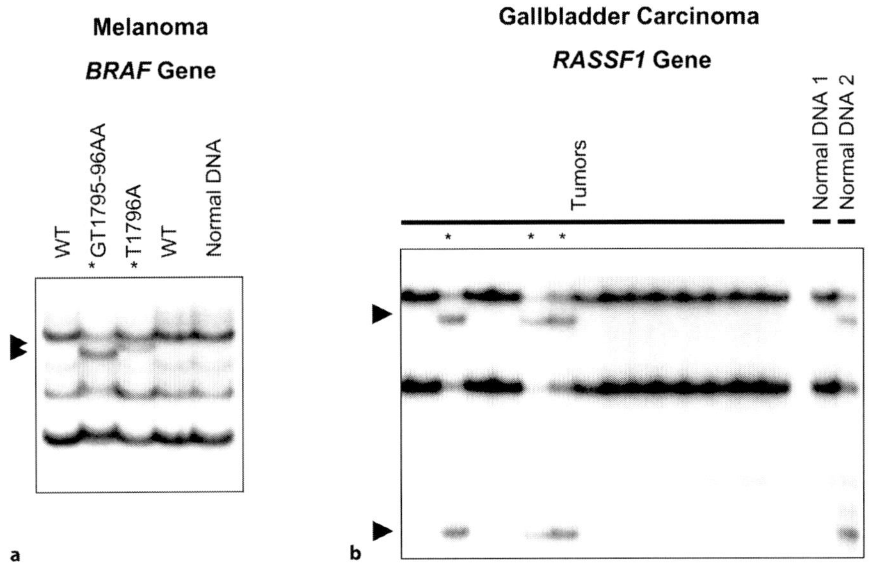

Fig. 8.2. SSCP examples for detection of mutations and polymorphisms in DNA extracted from FFPETs. **a** Exon 15 *BRAF* gene mutations in melanoma specimen and then confirmed sequence of the suspicious cases. **b** Tumor suppressor gene *RASSF1A* codon 133 polymorphism (GCT to TCT) in gallbladder carcinoma. Note that the suspicious bands are present in normal DNA 2. *Black arrow* indicates suspicious bands; * indicates confirmed changes in the sequencing analysis

The basic methodological considerations for SSCP analysis from FFPET are shown in the Protocols section (Chap. 11.5.1.5).

8.5.2 Restriction Fragment Length Polymorphism

This technique can be used to detect nucleotide changes by comparing the sizes of restriction endonuclease-cleaved genomic DNA fragments. For mutation analysis, PCR-amplified DNA sequences are subjected to RFLP. This methodology can be applied in cases in which the nucleotide change introduces an artificial restriction enzyme cleavage site or the particular mutation abolished an existing cleavage site (Franklin et al. 1997; Uribe et al. 2003). In the first situation, enzyme restriction digestion is seen in samples containing the mutant allele, and in the second in samples having the wild-type allele only. A representative example is shown in Fig. 8.3. In addition, a designed-RFLP method utilizing inner mismatched primers in the PCR amplification can be used to introduce a new restriction site into the PCR product derived from wild-type or mutant-type alleles, and the PCR product thus obtained is digested using a corresponding restriction en-

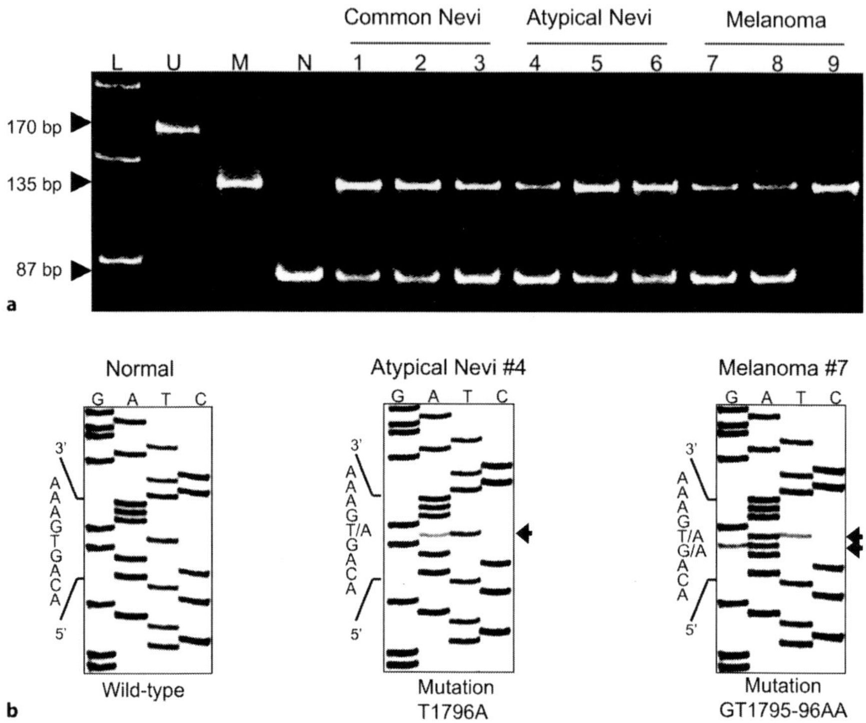

Fig. 8.3. RFLP mutation analysis of the oncogene *BRAF* at codon 599 in DNA obtained from microdissected FFPET melanocytic lesions of the skin. Mutation produces a loss of a restriction site for *Tsp*RI that can be detected by PCR-RFLP analysis. Three bands are shown: undigested, 170 bp; mutant, 135 bp; and wild type, 87 bp. Fragment lengths are given in bp on the *left*. *L* 50-bp DNA ladder, *U* uncut DNA, *M* mutated DNA control, *N* normal DNA control. Mutations were confirmed using manual sequencing methodology

zyme (Sugio et al. 1997). The most important methodological issues to consider for applying this methodology to FFPETs are shown in the Protocols section (Chap. 11.5.1.6).

8.6 Gene Methylation Analysis

Aberrant methylation (referred to as methylation) of normally unmethylated CpG islands in gene promoter regions is an important method for silencing tumor suppressor genes (TSGs; Herman and Baylin 2003). Methylation results in transcriptional inactivation of several TSGs in human cancer and serves as an alternative for the genetic loss of gene function by deletion or mutation. Development of new techniques that couple bisulfite modification with PCR has enabled these alterations to be studied in all

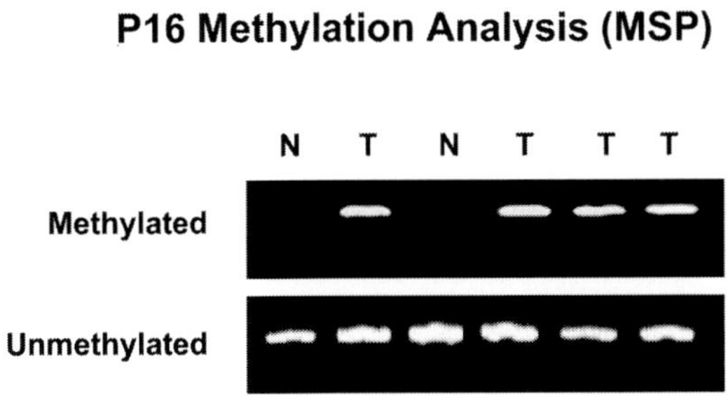

Fig. 8.4. Gene promoter methylation analysis using methylation-specific PCR (MSP) methodology from DNA extracted from FFPET from gallbladder carcinoma specimens. The amplified methylated and unmethylated allele products after step two of the nested MSP for the tumor suppressor gene *P16* are shown. *T* Tumor, *N* normal tissue

types of biological samples, including FFPETs (Herman et al. 1996). The methylation-specific PCR (MSP) method can rapidly assess the methylation status of virtually any group of CpG sites within a CpG island, independent of the use of methylation-sensitive restriction enzymes. This assay entails initial modification of DNA by sodium bisulfite, converting all unmethylated cytosines (but not the methylated cytosines, 5-methylcytosines) to uracil, and subsequent amplification with primers specific for methylated versus unmethylated DNA (Herman et al. 1996; Fig. 8.4). MSP requires only small quantities of DNA, is sensitive to 0.1% methylated alleles of a given CpG island locus, and can be performed on DNA extracted from FFPETs (Herman et al. 1996; Lee et al. 2002; House et al. 2003). MSP circumvents the false positive results inherent to previous PCR-based approaches, which relied on differential restriction enzyme cleavage to distinguish methylated from unmethylated DNA.

For MSP analysis using FFPET samples, see Protocols section (Chap. 11.5.1.7).

8.7 Clonality Analysis

8.7.1 Overview

Several techniques have been used to determine clonal association between tumors and lesions with the potential to develop into invasive tumors, namely microsatellite examination for LOH and MSI, X-chromosome inactivation, specific gene mutation and mitochondrial mutations analyses (Diaz-Cano et al. 2001). Specific technical information will be provided only for methodologies not discussed in other sections of this chapter.

8.7.2 Microsatellite Analysis

Microsatellite analysis measuring LOH and MSI have been used as clonal markers to determine the origins of synchronous or metachronous tumors and precursor lesions in several tumors (Tang et al. 2003). Because LOH frequencies are variable among chromosomal regions, and the informative (markers showing polymorphism) rates are also variable among markers, multiple markers must be used to gather the necessary information (Diaz-Cano et al. 2001; Tang et al. 2003). In addition, because there are three possible genetic patterns for any informative marker (i.e., retention, loss of upper allele, or loss of lower allele), about 33% of any paired lesions may exhibit an identical genetic pattern by chance alone. To increase confidence in the determination of clonal origin between two lesions, genetic patterns from multiple informative markers should be obtained. If two lesions share identical genetic patterns in a substantial number of markers, then they are likely to be derived from the same clone (Fig. 8.5A). However, a lack of shared genetic patterns in some markers between two lesions may occur even though the lesions derive from the same clonal origin because certain genetic changes may occur independently during tumor progression.

8.7.3 Gene Mutations

Specific gene mutations have been also used as genetic markers to distinguish the clonal relationship of multiple malignant and premalignant lesions (Diaz-Cano et al. 2001). The probability of the same clonal origin is very high if tumors share an identical particular mutation in a given gene. Some of those analyses have been based on mutations identified in tumor suppressor gene *TP53*, the most frequent oncogene mutated in human cancers (Harris 1996). However, there are two important limitations to this approach. Firstly, lesions need to contain the particular gene mutation, and secondly, some of those mutations occur relatively late in the tumorigenic process, therefore it may not be suitable for analyzing premalignant lesions or histologically normal appearing epithelium.

8.7.4 X-Chromosome Inactivation

The majority of investigators have used the human androgen receptor (*HUMARA*) assay for X-chromosome inactivation analysis (Mashal et al. 1993; Diaz-Cano et al. 2001; Jovanovic et al. 2003). There are two *Hha*I and two *Hpa*II restriction sites within the first exon of the X-linked *HUMARA* gene. They are located less than 100 bp 5′ from a highly polymorphic CAG microsatellite repeat. Methylation of these restriction sites is associated with X-chromosome inactivation in females. Therefore, digestion with methyla-

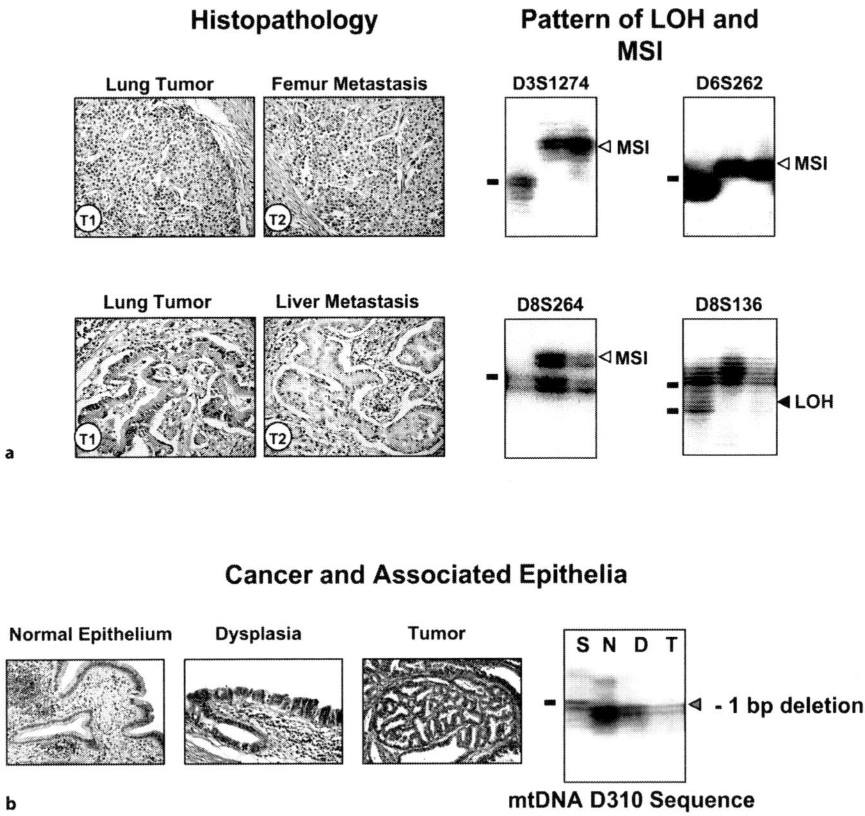

Fig. 8.5. Examples of clonality analysis in microdissected FFPET using microsatellite (**a**) and mitochondrial C-tract (D310) alterations (**b**). **a** Examples of microsatellite analysis (loss of heterozygosity, LOH, and microsatellite instability, MSI) for clonality in primary tumor and metastasis from autopsies. For each case, histopathology (*T1* primary tumor, *T2* metastasis), microsatellite abnormality pattern, and representative autoradiographs for LOH (*closed arrowhead*) and MSI (*open arrowhead*) are shown. Both cases showed identical or similar patterns of microsatellite abnormalities between primary tumor and metastasis, indicating a clonal relationship. *Horizontal bars* on the left of the autoradiographs indicate the main allelic bands. **b** Gel radiographs of mitochondrial C-tract (D310) alterations analysis by a length-based PCR assay run in a 6% polyacrylamide gel in DNA extracted from microdissected FFPET gallbladder cancer specimens. A case of gallbladder carcinoma with adjacent gallbladder epithelia showing the same D310 sequence deletion pattern indicating clonal relationship is shown. *S* stromal cells, *N* normal epithelium, *T* tumor, and *D* dysplasia. D310 sequence abnormalities were confirmed using manual sequencing methodology

tion-sensitive enzymes followed by PCR-amplification with primers flanking the sites can be used to distinguish between transcriptionally active and inactive X-chromosomes in heterozygous female subjects. In polyclonal tissues, one X-chromosome is active in some cells and the other in other cells. In contrast, in monoclonal tissues, all cells have inactivated the same X-chromosome. Therefore, if two lesions show hypermethylation in different X-chromosomes, then they must derive from different cell progenitors. If two lesions show an identical pattern of X chromosome hypermethylation, then there is a 50% chance that they have a common clonal origin. Although the *HUMARA* assay has been successfully used to determine clonality of clinical and research specimens, including FFPETs, this methodology is limited to female patients, and certain technical difficulties such as incomplete enzyme cleavage, inconsistent results and suboptimal PCR amplification frequently affect the interpretation of the results (Diaz-Cano et al. 2001).

The most important technical issues for X-chromosome inactivation analysis to assess clonality based on the *HUMARA* assay are (Jovanovic et al. 2003) analyzed in the Protocols section (Chap. 11.5.1.1.2).

8.7.5 Genotyping Assay of the D310 Repeat

Recently, a new marker has been added to the arsenal for determination of clonal outgrowth, the deletions/insertions examination at a polycytosine tract (C-tract) in displacement loop (D-loop) of human mitochondrial DNA (mtDNA), known as D310 sequence (Sanchez-Cespedes et al. 2001). Those abnormalities have been reported in a number of tumors (Sanchez-Cespedes et al. 2001) and in premalignant lesions associated to head and neck (Ha et al. 2002) and gallbladder carcinomas (Tang et al. 2003). Similar to the deletions/insertions measured by microsatellite analysis, the C-tract deletions/insertions appear to be good markers to determine clonal origin among multiple malignant and premalignant lesions (Fig. 8.5B). It will be interesting to see whether combining the C-tract and other clonality analyses can substantially increase the informative rate and provide sufficient information to determine the clonal origin of multiple cancers and premalignant lesions in most patients (Mao 2002). The most important protocol considerations for this type of analysis in DNA extracted from FFPETs are summarized in the Protocols section (Chap. 11.5.1.8).

8.8 Summary

The study of normal and neoplastic genomes has been the most widely used application of microdissected material in the study of cancer pathways. Genetic changes in the multistep progression of cancer can involve amplification or gain of function mutations in dominant oncogenes, or they may involve loss of function by deletion, mutation or methylation in recessive tumor suppressor genes. Microdissection has made a remarkable difference in the application of DNA analysis to the study of cancer pathways as well as allowing the study of sequential genetic changes in preneoplastic lesions. Although DNA extracted from formalin-fixed and paraffin-embedded tissues (FFPET) has been considered to be a poor substrate for molecular studies, several reliable methodologies for the examination of genetic abnormalities in FFPET-derived DNA have been well established and are discussed in this chapter. Their applications as well as the underlying causes of methodological problems and the potential techniques to overcome them are also reviewed.

References

Boland CR, Thibodeau SN, Hamilton SR, Sidransky D, Eshleman JR, Burt RW, Meltzer SJ, Rodriguez-Bigas MA, Fodde R, Ranzani GN, Srivastava S (1998) A National Cancer Institute Workshop on Microsatellite Instability for cancer detection and familial predisposition: development of international criteria for the determination of microsatellite instability in colorectal cancer. Cancer Res 58:5248–5257

Coombs NJ, Gough AC, Primrose JN (1999) Optimisation of DNA and RNA extraction from archival formalin-fixed tissue. Nucleic Acids Res 27:e12

de Lamballerie X, Chapel F, Vignoli C, Zandotti C (1994) Improved current methods for amplification of DNA from routinely processed liver tissue by PCR. J Clin Pathol 47:466–467

Diaz-Cano SJ, Brady SP (1997) DNA extraction from formalin-fixed, paraffin-embedded tissues: protein digestion as a limiting step for retrieval of high-quality DNA. Diagn Mol Pathol 6:342–346

Diaz-Cano SJ, Blanes A, Wolfe HJ (2001) PCR techniques for clonality assays. Diagn Mol Pathol 10:24–33

Franklin WA, Gazdar AF, Haney J, Wistuba II, La Rosa FG, Kennedy T, Ritchey DM, Miller YE (1997) Widely dispersed p53 mutation in respiratory epithelium. J Clin Invest 100:2133–2137

Frayling IM (2002) Methods of molecular analysis: mutation detection in solid tumours. Mol Pathol 55:73–79

Gillespie JW, Ahram M, Best CJ, Swalwell JI, Krizman DB, Petricoin EF, Liotta LA, Emmert-Buck MR (2001) The role of tissue microdissection in cancer research. Cancer J 7:32–39

Ha PK, Tong BC, Westra WH, Sanchez-Cespedes M, Parrella P, Zahurak M, Sidransky D, Califano JA (2002) Mitochondrial C-tract alteration in premalignant lesions of the head and neck: a marker for progression and clonal proliferation. Clin Cancer Res 8:2260–2265

Harris CC (1996) The 1995 Walter Hubert Lecture – molecular epidemiology of human cancer: insights from the mutational analysis of the p53 tumour-suppressor gene. Br J Cancer 73:261–269

Heinmoller E, Liu Q, Sun Y, Schlake G, Hill KA, Weiss LM, Sommer SS (2002) Toward efficient analysis of mutations in single cells from ethanol-fixed, paraffin-embedded, and immunohistochemically stained tissues. Lab Invest 82:443–453

Herman JG, Baylin SB (2003) Gene silencing in cancer in association with promoter hypermethylation. N Engl J Med 349:2042–2054

Herman JG, Graff JR, Myohanen S, Nelkin BD, Baylin SB (1996) Methylation-specific PCR: a novel PCR assay for methylation status of CpG islands. Proc Natl Acad Sci USA 93:9821–9826

House MG, Wistuba II, Argani P, Guo M, Schulick RD, Hruban RH, Herman JG, Maitra A (2003) Progression of gene hypermethylation in gallstone disease leading to gallbladder cancer. Ann Surg Oncol 10:882–889

Hung J, Kishimoto Y, Sugio K, Virmani A, McIntire DD, Minna JD, Gazdar AF (1995) Allele-specific chromosome 3p deletions occur at an early stage in the pathogenesis of lung carcinoma. JAMA 273:558–563

Jovanovic L, Delahunt B, McIver B, Eberhardt NL, Grebe SK (2003) Optimising restriction enzyme cleavage of DNA derived from archival histopathological samples: an improved HUMARA assay. Pathology 35:70–74

Knudson AG (1985) Hereditary cancer, oncogenes, and antioncogenes. Cancer Res 45:1437–1443

Lee S, Kim WH, Jung HY, Yang MH, Kang GH (2002) Aberrant CpG island methylation of multiple genes in intrahepatic cholangiocarcinoma. Am J Pathol 161:1015–1022

Legrand B, Mazancourt P, Durigon M, Khalifat V, Crainic K (2002) DNA genotyping of unbuffered formalin fixed paraffin embedded tissues. Forensic Sci Int 125:205–211

Maitra A, Wistuba II, Gazdar AF (2001) Microdissection and the study of cancer pathways. Curr Mol Med 1:153–162

Mao L (2002) A new marker determining clonal outgrowth. Clin Cancer Res 8:2021–2023

Mashal RD, Lester SC, Sklar J (1993) Clonal analysis by study of X chromosome inactivation in formalin-fixed paraffin-embedded tissue. Cancer Res 53:4676–4679

Paris PL, Albertson DG, Alers JC, Andaya A, Carroll P, Fridlyand J, Jain AN, Kamkar S, Kowbel D, Krijtenburg PJ, Pinkel D, Schroder FH, Vissers KJ, Watson VJ, Wildhagen MF, Collins C, Van Dekken H (2003) High-resolution analysis of paraffin-embedded and formalin-fixed prostate tumors using comparative genomic hybridization to genomic microarrays. Am J Pathol 162:763–770

Sanchez-Cespedes M, Parrella P, Nomoto S, Cohen D, Xiao Y, Esteller M, Jeronimo C, Jordan RC, Nicol T, Koch WM, Schoenberg M, Mazzarelli P, Fazio VM, Sidransky D (2001) Identification of a mononucleotide repeat as a major target for mitochondrial DNA alterations in human tumors. Cancer Res 61:7015–7019

Sato Y, Sugie R, Tsuchiya B, Kameya T, Natori M, Mukai K (2001) Comparison of the DNA extraction methods for polymerase chain reaction amplification from formalin-fixed and paraffin-embedded tissues. Diagn Mol Pathol 10:265–271

Savioz A, Blouin JL, Guidi S, Antonarakis SE, Bouras C (1997) A method for the extraction of genomic DNA from human brain tissue fixed and stored in formalin for many years. Acta Neuropathol (Berl) 93:408–413

Sugio K, Molberg K, Albores-Saavedra J, Virmani AK, Kishimoto Y, Gazdar AF (1997) K-ras mutations and allelic loss at 5q and 18q in the development of human pancreatic cancers. Int J Pancreatol 21:205–217

Taguchi T, Cheng GZ, Bell DW, Balsara B, Liu Z, Siegfried JM, Testa JR (1997) Combined chromosome microdissection and comparative genomic hybridization detect multiple sites of amplification DNA in a human lung carcinoma cell line. Genes Chromosomes Cancer 20:208–212

Tang M, Pires Y, Schultz M, Duarte I, Gallegos M, Wistuba II (2003) Microsatellite analysis of synchronous and metachronous tumors: a tool for double primary tumor and metastasis assessment. Diagn Mol Pathol 12:151–159

Tang M, Baez S, Diaz A, Calvo A, Pruyas M, Riquelme E, Wistuba II (2003) Mononucleotide sequence in the pathogenesis of gallbladder carcinoma. Clin Cancer Res 10:1041–1046

Uribe P, Wistuba II, Gonzalez S (2003) BRAF mutation: a frequent event in benign, atypical, and malignant melanocytic lesions of the skin. Am J Dermatopathol 25:365–370

Wang VW, Bell DA, Berkowitz RS, Mok SC (2001) Whole genome amplification and high-throughput allelotyping identified five distinct deletion regions on chromosomes 5 and 6 in microdissected early-stage ovarian tumors. Cancer Res 61:4169–4174

Wistuba II, Montellano FD, Milchgrub S, Virmani AK, Behrens C, Chen H, Ahmadian M, Nowak JA, Muller C, Minna JD, Gazdar AF (1997a) Deletions of chromosome 3p are frequent and early events in the pathogenesis of uterine cervical carcinoma. Cancer Res 57:3154–3158

Wistuba II, Lam S, Behrens C, Virmani AK, Fong KM, LeRiche J, Samet JM, Srivastava S, Minna JD, Gazdar AF (1997b) Molecular damage in the bronchial epithelium of current and former smokers. J Natl Cancer Inst 89:1366–1373

Wistuba II, Behrens C, Milchgrub S, Virmani AK, Jagirdar J, Thomas B, Ioachim HL, Litzky LA, Brambilla EM, Minna JD, Gazdar AF (1998) Comparison of molecular changes in lung cancers in HIV-positive and HIV-indeterminate subjects. JAMA 279:1554–1559

Wistuba II, Behrens C, Virmani AK, Milchgrub S, Syed S, Lam S, Mackay B, Minna JD, Gazdar AF (1999) Allelic losses at chromosome 8p21-23 are early and frequent events in the pathogenesis of lung cancer. Cancer Res 59:1973–1979

Wistuba II, Behrens C, Virmani AK, Mele G, Milchgrub S, Girard L, Fondon JW 3rd, Garner HR, McKay B, Latif F, Lerman MI, Lam S, Gazdar AF, Minna JD (2000) High resolution chromosome 3p allelotyping of human lung cancer and preneoplastic/preinvasive bronchial epithelium reveals multiple, discontinuous sites of 3p allele loss and three regions of frequent breakpoints. Cancer Res 60:1949–1960

Wistuba II, Tang M, Maitra A, Alvarez H, Troncoso P, Pimentel F, Gazdar AF (2001) Genome-wide allelotyping analysis reveals multiple sites of allelic loss in gallbladder carcinoma. Cancer Res 61:3795–3800

Wistuba II, Maitra A, Carrasco R, Tang M, Troncoso P, Minna JD, Gazdar AF (2002) High resolution chromosome 3p, 8p, 9q and 22q allelotyping analysis in the pathogenesis of gallbladder carcinoma. Br J Cancer 87:432–440

9 mRNA Applications

Fatih Ozsolak, Veronica Luzzi and Mark A. Watson

9.1 Introduction

Gene expression profiling has become an increasingly popular tool to survey the transcriptional landscape of human and animal tissues. This experimental approach has led to exciting new insights in biological signaling networks (Stuart et al. 2003) and advances in clinical diagnostics (Huang et al. 2003). However, the complex cellular heterogeneity of eukaryotic tissues such as human adenocarcinoma may often obscure significant patterns of gene transcription present in subpopulations of cells when homogenized tissue is considered in its entirety. If nothing else, this approach precludes a comparative analysis of expression profiles from independent cell populations within the same tissue specimen. Accordingly, laser capture microdissection (LCM) is an attractive tool to analyze gene expression profiles from microdissected cells in heterogeneous tissue sections (Leethanakul et al. 2000; Alevizos et al. 2001; Luzzi et al. 2001; Ma et al. 2003).

Use of LCM tissue for gene expression profiling poses three technical challenges. The first is the limited availability of properly preserved tissue specimens. Although less problematic for animal tissues, human clinical specimens are often preserved in formalin and paraffin wax, rendering RNA derived from these specimens generally unsuitable for current microarray technology. Second, the quantity of RNA derived from LCM tissue (often in the picogram to nanogram range) is insufficient for standard microarray target preparation, thus necessitating the use of transcript amplification. Finally, production of quality microarray data is highly dependent upon a high-quality RNA template and requires stringent quality assurance measures.

In this chapter, we will provide a theoretical overview and sample protocols for performing RNA isolation, transcript amplification, microarray analysis, and RT-PCR confirmation of gene expression profiles from LCM tissue. Figure 9.1 provides a generalized flow diagram of this approach.

Principles and Practice
Emmert-Buck et al. (Eds.):
Dissecting the Molecular Anatomy of Tissue

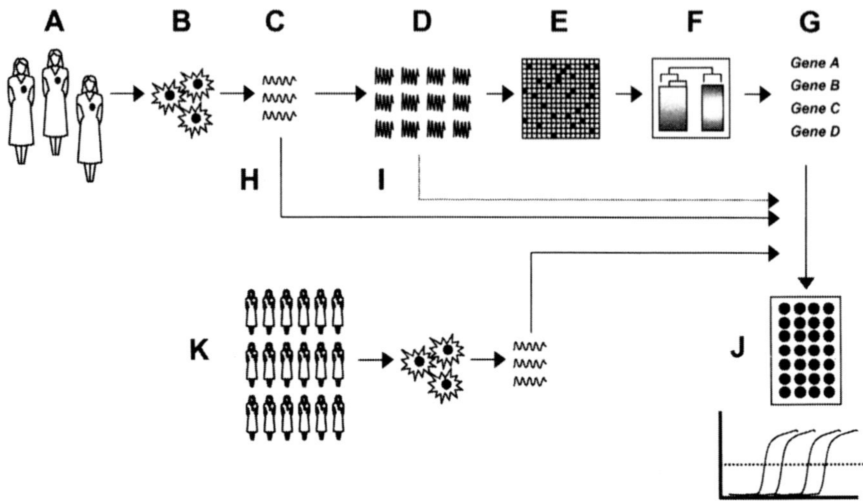

Fig. 9.1. Experimental approach for microarray analysis and RT-PCR validation of gene expression from LCM tissue. A set of reasonably accessible fresh frozen clinical specimens (**A**) are used for LCM to isolate cell populations of interest (**B**). Nanogram quantities of RNA are isolated from dissected cells (**C**), which is then quantified and qualitatively assessed for integrity. Approximately 10–30 ng of RNA is amplified using T7 RNA polymerase-based in vitro transcription (**D**) and labeled cRNA is applied to the microarray (**E**). After bioinformatic analysis of gene expression patterns (**F**), a set of candidate sequences are chosen for further analysis (**G**). Expression patterns are validated using qRT-PCR (**J**) with either the originally isolated RNA sample (**H**) or an aliquot of the amplified RNA (**I**). Additional specimens, often derived from FFPE tissue blocks (**K**), can then be used to further validate or expand patterns of gene expression identified in the pilot sample set

Although these methods are applicable to all animal tissues, we will focus on the analysis of human clinical specimens, which can be particularly challenging. Examples of microarray analysis will be based on the use of Affymetrix GeneChip microarrays, although most protocols and guidelines presented are equally applicable to other microarray platforms. In addition to practical considerations and protocols, a number of troubleshooting guidelines based on our own experience will be provided.

9.2 Tissue Selection for Gene Expression Analysis

Previous chapters in this volume as well as other sources (Huang et al. 2002) have described the critical parameters necessary for optimized tissue processing for LCM. For gene expression analysis using nucleic acid microarrays, fresh frozen tissue remains the gold standard from which to isolate RNA. Frozen tissue may be more difficult to procure and store, provides poorer histological detail, and requires considerable care to isolate high quality RNA. However, until more robust methods are available for performing microarray experiments on RNA derived from formalin-fixed, paraffin-embedded (FFPE) tissues (Karsten et al. 2002), fresh frozen tissue is a necessity.

On the other hand, clinical tissue specimens embedded in paraffin are plentiful (although it may be increasingly difficult to request these specimens from surgical pathology departments). The histological detail of FFPE tissue is obviously excellent and these tissues are easy to work with. Although the RNA isolated from FFPE tissue is severely degraded and/or chemically modified (Masuda et al. 1999) making it unsuitable for most microarray platforms, it can be successfully utilized for quantitative RT-PCR (qRT-PCR) assays with appropriately designed primers (Godfrey et al. 2000; Specht et al. 2001).

Based on these constraints, a reasonable experimental approach is to perform LCM on a limited number of fresh frozen tissue samples for microarray analysis, identify candidate transcripts of interest, and then validate gene expression patterns using qRT-PCR and a larger number of RNA samples derived from microdissected FFPE tissue (Fig. 9.1).

9.3 RNA Isolation

9.3.1 Overview

Protocols for the isolation of RNA from frozen LCM tissue, as well as from FFPE LCM tissue, are shown in the Protocols section (Chap. 11.4.2.2). There are generally two methodological approaches for RNA isolation from LCM tissue. The first involves differential binding and elution of cell lysates to microspin columns. Several commercial reagent systems are available from vendors such as Arcturus (Mountain View, CA), Ambion (Austin, TX), and Stratagene (LaJolla, CA). The second approach involves organic extraction and precipitation using a scaled-down version of traditional RNA isolation methods. The reagents necessary for this protocol can be purchased independently and are also packaged as commercial reagent systems from vendors such as Ambion and Stratagene. Spin column protocols are fast and convenient, and avoid the use of toxic chemicals such as phenol and chloroform. They may be particularly efficient when the yield of RNA is expected to be very low (< 10 ng). However, in our experience, the organic extraction and precipitation method more consistently yields high quality RNA, particularly when tissue is collected from more than 1000 cells (i.e., an anticipated RNA yield of > 10 ng). The protocol presented in the Protocols section (Chap. 11.4.2.2) is routinely used in our laboratory to isolate high quality RNA from frozen LCM tissue. A protocol for the isolation of RNA from FFPE LCM tissue is presented in Chapter 11.4.2.3.

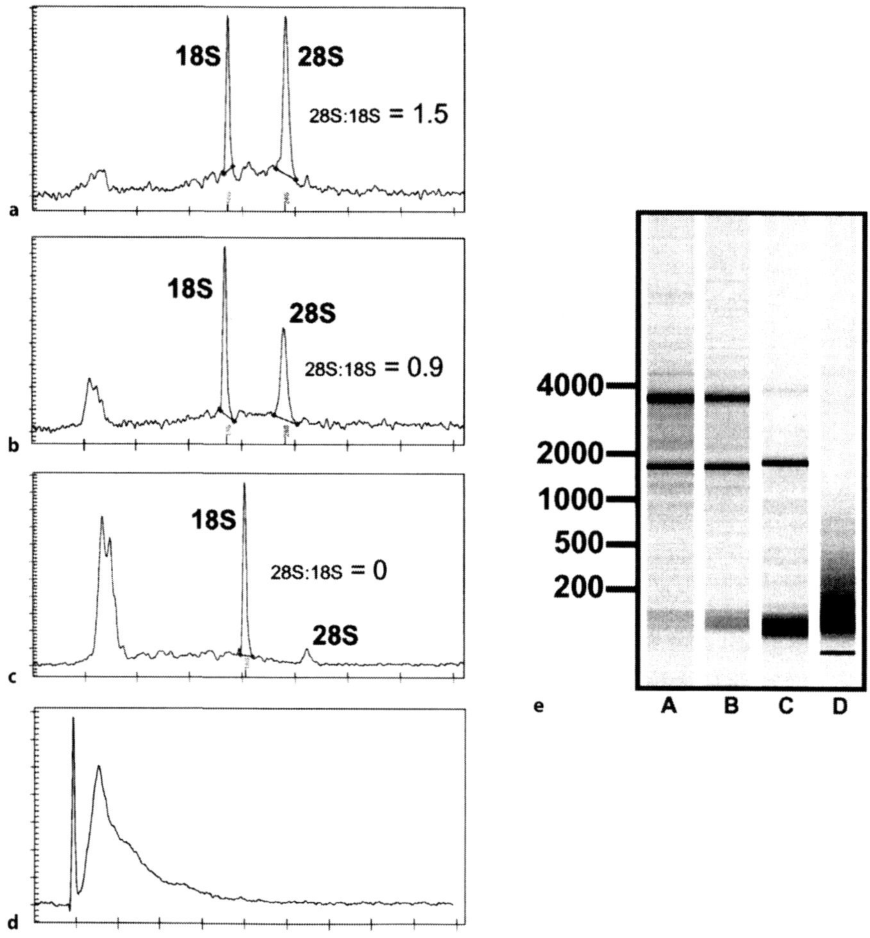

Fig. 9.2. Qualitative analysis of RNA derived from LCM of frozen or FFPE tissue. RNA isolated from fresh frozen tissue (**a–c**) or FFPE tissue (**d**) was analyzed using the Agilent Bioanalyzer 2100 and RNA Nano Chip Assay. The graphs (**a–d**) demonstrate fluorescence intensity as a function of migration time. Peaks corresponding to 28S and 18S ribosomal bands along with their calculated band intensity ratios are noted. **a** High quality RNA; **b** moderate quality RNA; **c** severely degraded RNA; **d** RNA derived from FFPE tissue. Samples **a** and **b** were used for transcript amplification and produced microarray data of comparable quality (see Table 9.1, samples 14 and 15). Note the lack of a 28S ribosomal band and increased intensity of low molecular weight RNA in degraded sample **c**. Distinct ribosomal bands are not present in RNA derived from FFPE LCM tissue, only a low molecular weight nucleic acid smear with a mean size of 100–150 nucleotides. The sharp low molecular peak is a reference standard. **E** Pseudoelectrophoretogram of RNA samples **a–d**; molecular weight markers are indicated

9.3.2 Expected Results

Figure 9.2 illustrates typical pseudoelectrophoretograms obtained with RNA samples isolated from fresh frozen and FFPE tissues by LCM. Ideally, the intensity ratio of 28S to 18S ribosomal RNA bands should be 2.0 in intact RNA preparations derived from frozen tissue. However, this is seldom

Table 9.1. Representative yields and quality of cellular RNA isolated from LCM of fresh frozen breast epithelial tissue (tumor and nonmalignant cells). The number of 30-μm laser shots performed for each dissection is indicated; 28S/18S, ratio of ribosomal RNA band intensities determined by Agilent RNA 6000 Nano Chip. Resulting yields of biotin-labeled cRNA synthesized from 20 ng of each total cellular RNA, and corresponding quality of GeneChip microarray data derived from each cRNA hybridization are also indicated. %P, Percentage of microarray probes with a signal that was statistically above background, as determined by Affymetrix Microarray Analysis Suite 5.0 software. All targets were hybridized to Affymetrix U95A human GeneChip microarrays, representing approximately 12,600 gene transcripts. Note that RNAs with 28S/18S ribosomal ratios of less than 0.8 yield low amounts of cRNA and poor results from microarray hybridization (samples 1, 2). In rare cases, relatively high yields of cRNA may still result in poor results after microarray hybridization (sample 3)

Sample	No. of shots	RNA yield (ng)	28S/18S	cRNA yield (μg)	%P
1	1700	30	0.7	18	14
2	2454	74	0.3	18	15
3	3467	40	1.2	40	17
4	3245	190	1.6	18	19
5	1100	40	0.2	18	23
6	4323	130	0.9	21	24
7	1097	45	1.0	20	27
8	2800	190	1.2	20	27
9	4212	100	0.9	18	29
10	5425	220	1.0	30	30
11	1700	70	0.9	18	31
12	3000	180	1.1	40	31
13	2500	130	1.3	18	33
14	3700	150	1.5	40	33
15	3400	110	0.9	24	34
16	2750	140	1.0	30	34
17	3245	190	1.6	18	35
18	1428	55	1.2	20	36
19	1500	110	0.8	24	36
20	1552	120	1.0	23	36
21	2357	40	0.8	18	38
22	3100	130	1.2	40	38
23	3300	100	1.3	40	39

the case. Ribosomal 28S/18S band ratios may be as low as 1.0, and RNA may still yield satisfactory microarray data (Table 9.1). Of greater importance is the presence or absence of a low molecular weight 'smear' below the 18S band. The presence of such a smear suggests severe RNA degradation. Moreover, the complete absence of a 28S ribosomal band is indicative of degraded RNA that will not perform well in microarray analysis. On the other hand, RNA isolated from FFPE LCM tissue almost always appears as a low molecular weight smear, with a mean size of 100–200 nucleotides.

When suspended in nuclease-free water, RNA samples with a UV 260/280 absorbance ratio of less than 1.6 are likely contaminated with trace organics or protein, even if their electrophoresis profile appears acceptable. This is particularly problematic when RNA is isolated by organic extraction rather than spin-column-based protocols. Trace organics or protein contaminants can inhibit downstream enzyme reactions. Therefore, if measured 260/280 ratios are significantly lower than 1.6, RNA should be re-precipitated.

Table 9.1 summarizes representative yields of RNA obtained from LCM of fresh frozen tissue. Depending on the tissue and cell type dissected, 1000 laser pulses with a 30-μm beam should yield 20–60 ng of total RNA. Yields of RNA from FFPE tissues are generally comparable. Note that estimation of RNA yield by electrophoresis and ribosomal band intensities may be as much as twofold discrepant with calculated yields based on UV absorbance or dye binding assays.

9.3.3 Troubleshooting

If RNA isolated from frozen LCM tissue is degraded or undetectable, it is useful to examine RNA isolated from the tissue section remaining after LCM (the 'tissue control'). If the quality of the tissue control RNA is acceptable, this suggests that an insufficient amount of tissue was dissected or that RNA was lost or degraded during the LCM process. If RNA isolated from the tissue control is also degraded, consider whether the tissue was properly preserved (i.e., snap frozen and maintained in a frozen state), the staining procedure was not performed correctly, or RNA isolation was not performed correctly. A single fresh frozen section from the tissue block may be cut and placed immediately into lysis buffer. If the resulting RNA from this tissue section is also degraded, this suggests a problem with tissue preservation or RNA isolation procedure. Consider using freshly isolated tissue or performing a control RNA isolation using tissue culture cells. If RNA from the tissue block is of high quality, but post-LCM tissue RNA is degraded, the tissue staining protocol should be re-evaluated. Make certain that stains compatible with RNA retrieval are used (Huang et al. 2002).

9.4 Transcript Amplification

9.4.1 Overview

Protocols for transcript amplification, cDNA synthesis and purification (round 1), in vitro transcription and purification (round 1), cDNA synthesis and purification (round 2), and biotin labeled in vitro transcription and purification (round 2) are shown in the Protocols section (Chap. 11.5.2.4). Because the quantity of RNA isolated from LCM (0.1–100 ng) is at least an order of magnitude less than that required for microarray target labeling (1–10 μg), amplification of LCM-derived RNA is necessary for most microarray platforms. RNA amplification may be performed using PCR-based (exponential) or in vitro transcription-based (linear) methods. Although PCR-based methods were originally thought to introduce significant amplification bias, newly modified protocols have demonstrated that this approach can result in accurate representation of transcript pools and have been used successfully in a number of recent gene expression profiling studies (Iscove et al. 2002; Chiang and Melton 2003). Nonetheless, use of T7 RNA Polymerase mediated in vitro transcription as originally described by Van Gelder et al. (1990) remains the most popular approach for transcript amplification. Commercial reagent systems from vendors such as Arcturus (Mountain View, CA), Ambion (Austin, TX), and Agilent (Palo Alto, CA) are readily available, reliable, and easy to use. For example, we routinely use the RiboAmp reagent system (Arcturus) to generate amplified targets (Luzzi et al. 2003).

The principle of T7 RNA polymerase amplification is outlined in Fig. 9.3. Total RNA is converted into first strand cDNA using a dT-T7 RNA polymerase promoter primer. A double-stranded cDNA is then created using either traditional Gubler-Hoffman second-strand synthesis or by priming with randomers. The double-stranded cDNA pool serves as a template for in vitro transcription, where T7 RNA polymerase creates a several 1000-fold amplification of antisense (aRNA) or complementary (cRNA) transcripts corresponding to the original cellular mRNA pool. This process is repeated for a second round of amplification, creating first strand cDNA using randomer priming of the aRNA, followed by second-strand synthesis with the dT-T7 promoter primer, and then in vitro transcription using T7 RNA polymerase. The level of amplification can be adjusted by performing a variable number of rounds of cDNA synthesis and in vitro transcription. Starting with 10–50 ng of cellular RNA, two rounds of amplification generally produce a sufficient quantity of target for array hybridization. This approach can be adapted to any microarray platform by incorporating direct or indirect labels at the appropriate step in the protocol during in vitro transcription for RNA-based targets and during cDNA synthesis for cDNA-based targets. A protocol for transcript amplification is presented in the Protocols Section (Chap. 11.5.2.4).

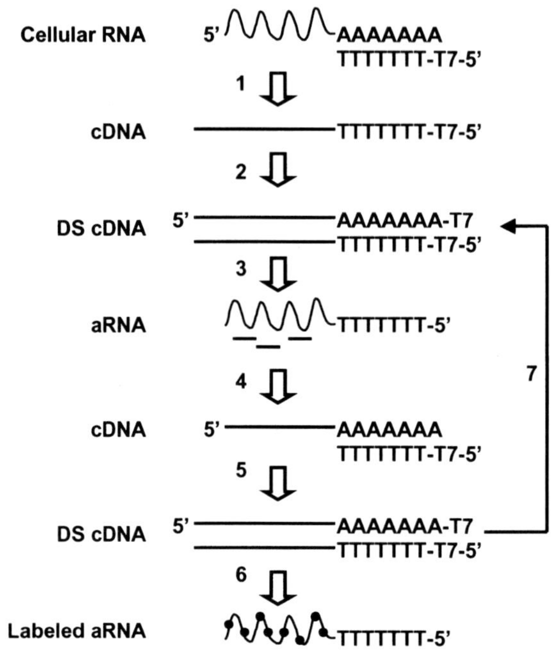

Fig. 9.3. Outline of linear amplification procedure. Cellular RNA is converted into first strand cDNA using an oligo-dT/T7 RNA polymerase promoter primer (*1*) and double-stranded (DS) cDNA is then made using standard Gubler-Hoffman synthesis (*2*). The resulting cDNA template is used for in vitro transcription to generate cRNA (*3*). Amplified cRNA is converted back into first strand cDNA using random primers (*4*) and then into double-stranded DNA using the dT/T7 primer (*5*). The amplified cDNA can be used again for in vitro transcription to generate labeled cRNA target for microarray hybridization (*6*). Alternatively, another round of transcript amplification may be performed (7)

9.4.2 Expected Results

Table 9.1 provides representative data of cRNA yields, starting with 20 ng of total RNA from microdissected human breast epithelium. Using the protocol provided above, approximately 20 μg of biotin-labeled cRNA should be generated for every 10 ng of input RNA. Significantly lower yields than this suggest amplification failure. However, the total yield of nucleic acid is not always an indicator of successful target amplification. In some cases, perfectly acceptable yields of labeled cRNA have produced low signal intensities upon microarray hybridization (Table 9.1). This phenomenon may be due to spurious amplification of non-mRNA templates or amplification of truncated (degraded) mRNA. Since the protocol results in truncated cDNAs subsequent to first strand cDNA synthesis with randomers (Fig. 9.3, step 4), the average size of labeled aRNA is slightly smaller after two rounds of

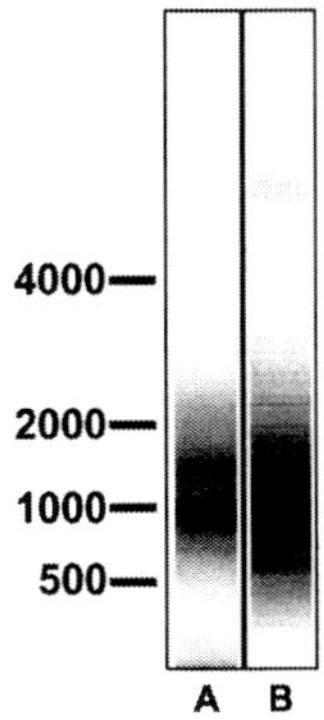

Fig. 9.4. Qualitative analysis of amplified cRNA. In *lane A*, 5 μg of total cellular RNA was utilized with a single round of transcript amplification to generate labeled cRNA. In *lane B*, 20 ng of the same total cellular RNA was utilized with two rounds of transcript amplification to generate labeled cRNA. A 1-μl aliquot of each cRNA was analyzed using an Agilent Bioanalyzer 2100 and RNA Nano Chip assay. Molecular weight markers are indicated. Note the similarity in size distribution of the cRNA populations, with cRNA generated from two rounds of amplification being slightly shorter

amplification as compared to a single round (Fig. 9.4). If microarray probes (e.g., spotted oligonucleotides) are not biased toward the 3′ end of their intended transcripts, this can result in a diminished number of positive probe signals. This is rarely the case for GeneChip microarrays and full length spotted cDNA microarrays.

When technical replicates are performed, even from independent dissections of the same tissue, reproducibility should be comparable to traditional target synthesis methods (Luzzi et al. 2003). A representative example of this reproducibility is illustrated in Fig. 9.5. However, compared to standard approaches, performing subsequent rounds of transcript amplification does introduce bias into the transcript pool (Fig. 9.5). For this reason, it is highly recommended that all targets for a given experiment be prepared in parallel using the same methodology.

9.4.3 Troubleshooting

Initially, 10 ng of control RNA (consisting of a high quality RNA source diluted to 10 ng/μl) should be utilized with all experimental samples for target amplification and microarray analysis. Successful amplification and hybridization of the control RNA, but only a few or none of the experimental RNAs suggests problems with RNA isolation. Qualitatively re-assess the RNA by any of the methods discussed in Chapter 11.4.2.2. If neither experimental nor control RNAs yield successful results, the reagent most commonly associated with failed target syntheses is the $T7T_{24}$ oligonucleotide primer. Any degradation or truncated synthesis of the promoter primer can destroy the T7 RNA polymerase site and result in cDNA templates with nonfunctional transcription promoters. The primer should be HPLC- or gel-purified to ensure that only full-length primer is used. Fresh primer should be utilized and primer stocks should be aliquoted to avoid repeated freeze–thaw cycles. The second most common source of systematic error is

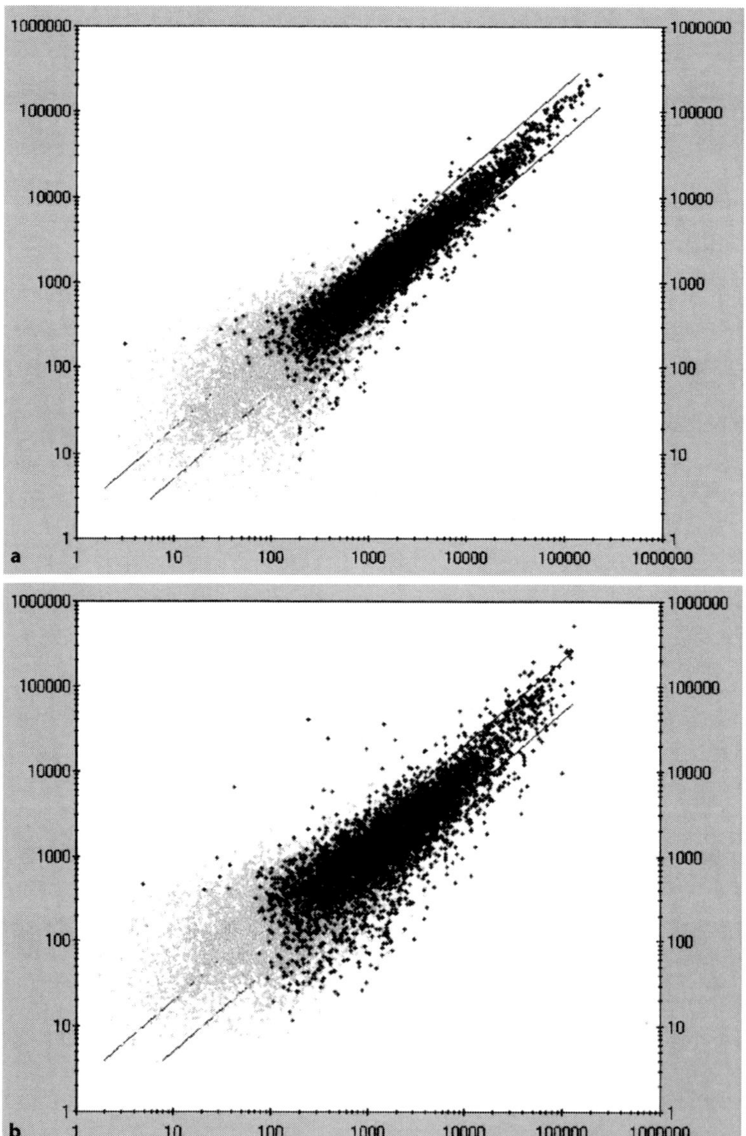

Fig. 9.5 a, b

carry over of ammonium acetate salts from precipitation into subsequent enzyme reactions. Be certain to centrifuge samples at room temperature and perform thorough washings with 70% ethanol prior to drying the nucleic acid pellets. Conversely, because pelleted nucleic acid is often invisible, extra care should be exercised to avoid washing away pellets.

9.5 RT-PCR Validation

9.5.1 Overview

Protocols for the RT-PCR validation are shown in the Protocols section (Chap. 11.5.2.5). While gene expression microarrays are an excellent discovery tool, their relative expense and requirement for high quality RNA make them less attractive validation assays. Since most microarray data sets consist of large numbers of variables (genes) with fewer numbers of observations (samples), independent validation of candidate gene expression profiles identified by microarray analysis is almost mandatory. Quantitative RT-PCR (qRT-PCR) is a convenient means by which to perform validation. Compared to frequently unavailable or poorly performing antibodies used for immunohistochemical studies, optimized gene-specific oligonucleotide primers can be readily designed and synthesized cheaply. As its name implies, the assay is quantitative (with at least 1.5- to 2.0-fold precision) and it can be performed in a relatively high-throughput fashion. Finally, qRT-PCR is particularly suited for the analysis of small quantities of often-fragmented RNA, such as that routinely obtained from LCM of FFPE tissue blocks.

Fig. 9.5. Reproducibility and comparability of amplified RNA samples. **a** Two independent LCM dissections were performed on fresh frozen human breast tissue to isolate the same population of nonmalignant epithelial cells from serial tissue sections. RNA was extracted from each LCM specimen independently and 20 ng of each RNA was utilized for transcript amplification and biotin labeling as described in this chapter. Biotinylated targets were applied to Affymetrix U95A GeneChip microarrays, and the scaled signal intensity from the two replicate samples are represented as a scatter plot, where transcript signal intensity in one sample is plotted on the *x-axis* and signal intensity for the same transcript in the second sample is plotted on the *y-axis*. *Black points* represent transcripts with signals that were statistically above background in one or both samples ('P' calls). *Gray points* represent transcripts with signals that were not statistically above background in either sample ('A' calls). *Diagonal lines* represent limits of twofold difference in signal between the two samples. **b** In this experiment, 5 μg of RNA was converted into biotinylated cRNA using a single round of in vitro transcription while 10 ng of the same RNA was converted into biotinylated cRNA using two rounds of in vitro transcription. An equal amount of biotinylated cRNA was hybridized to each of two Affymetrix U95A GeneChip microarrays. Data are represented as described in **a**

"Validation" may consist of two independent or parallel processes. First, the same RNAs utilized for initial microarray studies may be used to confirm patterns of gene expression with qRT-PCR ("technical validation"). For this purpose, we routinely attempt to isolate 50–100 ng of RNA from LCM of fresh frozen tissue. Approximately 10–30 ng of RNA is used for microarray analysis and the remainder can be used for this first set of validation studies (Fig. 9.1 H). If cellular RNA is limiting, it may also be possible to reliably utilize amplified cRNA (Fig. 9.1 I) for validation studies (Kabbarah et al. 2003; Ma et al. 2003). A second, larger independent set of RNAs not utilized for microarray analysis may also be used with qRT-PCR to extend or refute expression patterns identified in the pilot set of samples ("biological validation", Fig. 9.1 K). Often, this sample set may consist of archived FFPE blocks from the surgical pathology department.

The qRT-PCR method is performed by converting LCM-isolated RNA (either from frozen or FFPE tissue) into first strand cDNA using either an oligo-dT, randomer, or gene-specific primer. The first strand cDNA is then used as a template for PCR, where accumulation of the gene-specific amplification product is monitored by direct or indirect fluorescent detection. With each subsequent PCR cycle, the amount of amplification product increases and so does the fluorescence. At some cycle number, fluorescent intensity exceeds an arbitrarily defined threshold. This is referred to as the cycle threshold (C_T). The number of PCR cycles required to reach this threshold is inversely proportional to the starting amount of gene-specific cDNA (and hence RNA) template present in the reaction. By comparing normalized C_T values between samples, the relative (or absolute) abundance of original RNA present in the sample can be calculated (Bustin 2002).

9.5.2 PCR Assay Design

For the uninitiated, there is a plethora of options for designing qRT-PCR validation assays. One consideration is whether to employ "one-step" or "two-step" enzymology. In the single step protocol, RNA is converted to cDNA (usually with a gene-specific primer) and the cDNA is then amplified in the same tube. This approach is more convenient, but offers less flexibility. In the two-step protocol, RNA is converted to cDNA with oligo-dT or random hexamers. The resulting cDNA is then partitioned into several different PCR reactions, each employing a different set of gene-specific amplification primers. In this manner, PCR assays for each transcript are performed on the same common pool of cDNA.

There are also two basic fluorescent chemistries that may be employed in qRT-PCR assays. SybrGreen is a dye that fluoresces when bound to double stranded DNA. When added to a PCR reaction, increasing SybrGreen fluorescence is proportional to accumulation of double stranded amplification product, presumably that of the specific gene transcript to which amplification primers have been designed (Morrison et al. 1998). The advantage of the SybrGreen approach is that it is relatively inexpensive, requiring only one pair of standard amplification primers. Using individual reaction components, assays can be performed for less than 50 cents per reaction. The disadvantage of this method is that SybrGreen lacks sequence-binding specificity such that fluorescence from nonspecific amplification and primer dimer products is indistinguishable from that generated by the sequence-specific amplicon. This translates into a generally lower signal to noise ratio (particularly when using small amounts of input RNA) and a need to carefully validate primers before using them.

An alternative to using SybrGreen dye is the use of sequence-specific fluorescent probes. In this approach, transcript amplification occurs through the use of a pair of standard PCR primers, but a third fluorescent probe is required to bind to the amplified product to generate a fluorescent signal. This signal may be generated by polymerase cleavage of a fluorescent quencher (i.e., Applied Biosystems' TaqMan assay) or through a conformational change of the probe upon binding to the amplified product (i.e., Molecular Beacons). In either case, the use of a third gene-specific sequence greatly increases the specificity of the assay. Furthermore, pre-validated primer/probe sets for many human and mouse transcripts are commercially available (e.g., Applied Biosystems' Assay-on-Demand program), making assay set-up quick and convenient. The main disadvantage of this approach is the expense associated with fluorescent probes. When attempting to validate the expression pattern of 15–30 independent transcripts identified through microarray studies, the cost is often prohibitive. Furthermore, in our experience, while commercially prepared primer/probe sets work robustly on nanogram quantities of RNA, they often fail to perform with RNA isolated from LCM of FFPE tissue. A likely explanation for this anecdotal result is that the amplicon sizes of these assays exceed the average length of cDNA generated from the highly fragmented, FFPE-derived RNA.

No matter which of the several enzyme chemistries discussed above are employed, many vendors such as Applied Biosystems, Qiagen, Invitrogen, and Stratagene market pre-formulated enzyme mixes to which are added cDNA template and primer. Although convenient, these commercial reagents can cost two- to threefold more (US$ 0.85–1.25) per reaction, an important consideration when attempting to validate the expression of many genes on a large number of samples.

9.5.3 Quantification

There are three basic approaches to quantifying gene expression data from qRT-PCR assays. The absolute copy number method uses a synthetic RNA standard corresponding to the gene of interest (such as that created from in vitro transcription of a cDNA expression plasmid) at defined copy numbers. This RNA is used to create a standard curve that relates absolute number of RNA molecules to C_T number and that can be used to interpolate the number of mRNA copies in each experimental sample. While this approach provides the most definitive quantitative data, it is extremely labor-intensive to create RNA standards for each transcript to be assayed, particularly when many transcripts are to be examined.

A second approach relies on the generation of a relative standard curve. Here, a reference cellular RNA is serially diluted, reverse transcribed, and amplified to create a standard curve that relates fold dilution of input RNA to C_T number and that can be used to interpolate the relative fold difference in transcript abundance between samples. This method is considerably simpler than the absolute copy number method, but still requires a set of standard curve reactions to be performed along with every set of experimental samples.

The simplest approach to quantifying data uses the $\Delta\Delta C_T$ calculation (Livak and Schmittgen 2001). In this approach, the C_T of the target gene in the experimental sample is normalized to that of a control gene in the experimental sample, and then expressed relative to this difference in a baseline sample. The appeal of this method is that a standard curve is not necessary to express relative changes in gene expression between samples. However, an important caveat of this approach is that the efficiency of target gene and control gene amplification must be comparable (Fig. 9.6).

9.5.4 Normalization Controls

An important aspect of qRT-PCR validation studies is the selection of an appropriate "housekeeping gene" with which to normalize the expression levels of target genes across samples. Practically, the purpose of a housekeeping gene is to compensate for differences in RNA concentration and enzyme reaction efficiency between samples and therefore, the expression of the housekeeping gene should be invariant between all samples in the experiment. Glyceraldehyde-3-phosphate dehydrogenase (GAPDH), 18S ribosomal RNA (18S), beta-glucuronidase (GUSB), and beta-actin (ACTB) have all been used as housekeeping genes and several reports have published compendiums of other suitable control genes (Hamalainen et al. 2001; Vandesompele et al. 2002;). In reality, the selection of an appropriate control gene is entirely dependent upon the experimental paradigm. While GAPDH may be a suitable normalizing transcript in some studies, expression levels can vary greatly between nonmalignant cells and more metabolically active microdissected tumor cells. In practice, if qRT-PCR is being performed to validate microarray results, it is often useful to use the microarray data itself to identify a highly and invariantly expressed transcript in the experimental sample set. Assuming that the set of validation samples will be of similar composition, such transcripts identified within the mi-

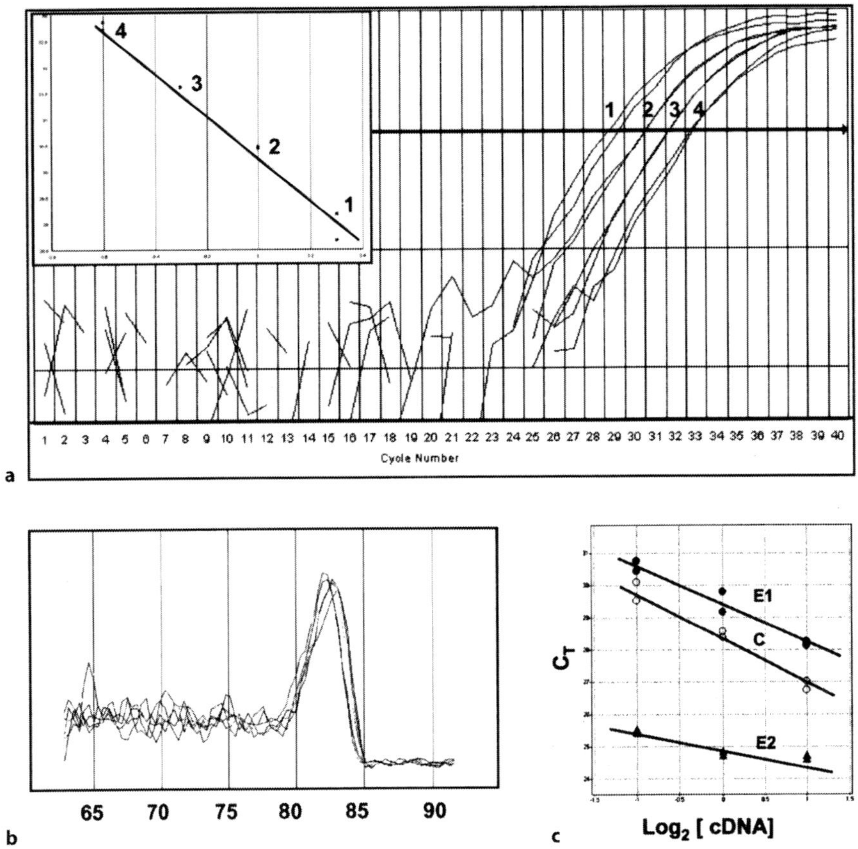

Fig. 9.6. Validation of primer pairs for qRT-PCR. **a** PCR amplification plot obtained from serial dilutions of cDNA standard, performed in duplicate. The ΔRn of SybrGreen fluorescence is plotted as a function of cycle number. The *horizontal line* intersecting the four pairs of amplification plots indicates the defined fluorescence threshold. Note a linear shift in the amplification curves as replicate samples of cDNA containing the equivalent input RNA of 20 ng (*1*), 10 ng (*2*), 5 ng (*3*), and 2.5 ng (*4*) are amplified. A plot of log [cDNA] vs. C_T (*inset*) demonstrates this relationship. **b** Melt curve analysis of the amplicons generated in **a**. The derivative of SybrGreen fluorescence is plotted as a function of increasing temperature. Note a single and specific change in fluorescence at $\approx 83\,^{\circ}$C, consistent with a single, uniform amplification product. **c** Plots of log [cDNA] vs. C_T using a primer pair for a normalizing control gene, "*C*", and two experimental gene primer pairs, "*E1*" and "*E2*". Each primer pair was tested with a serial dilution of control cDNA as described in **a**. Note that the slopes of the plots for control and E1 primer pairs are comparable, making the use of the $\Delta\Delta C_T$ quantification method applicable. The slope of the plot for E2, however, is sufficiently different so that the $\Delta\Delta C_T$ method should not be used

croarray data should serve as suitable controls for qRT-PCR validation. A protocol for RT-PCR validation is presented in the Protocols section (Chap. 11.5.2.5).

9.5.5 Expected Results

Figure 9.6 demonstrates typical results of a primer validation experiment performed using SybrGreen chemistry on an Applied Biosystems SDS7000 instrument. Duplicate reactions should be less than a single C_T apart. A plot of log [input cDNA] vs. C_T for twofold serial dilutions of control cDNA should be linear. In addition, if the $\Delta\Delta C_T$ method is to be used for relative quantification, the slopes of both target sequence and "housekeeping" normalizing control sequence should be parallel. The C_T values for samples at all dilutions should be less than 35 cycles. Melting curve analysis should indicate a uniform amplification product. In addition, template from the cDNA synthesis reaction in the absence of reverse transcriptase (i.e., negative control reaction) should result in an amplification plot that never reaches the defined fluorescence threshold and/or is suggestive of only a nonspecific amplification product upon melting curve analysis. Documentation of these features suggests that the primer sequences are sufficiently robust for quantitative analysis of the experimental LCM specimens.

When gene expression patterns defined by robust microarray data are selected for technical validation by qRT-PCR, results should be at least

Table 9.2. Representative fold-change values in gene expression between preinvasive (DCIS) and invasive breast tumor cells obtained from LCM of six different frozen human breast adenocarcinoma tissue specimens. The same RNA was used for both transcript amplification (30 ng) followed by U95A GeneChip microarray analysis, and cDNA synthesis (50 ng) followed by qRT-PCR analysis to validate the expression pattern of three genes. Fold change in gene expression between the two different dissected cell populations in each tumor was calculated using Affymetrix Microarray Analysis Suite 5.0 software for microarray data and the DDC_T method (see text) for qRT-PCR data. Note the general qualitative concordance between methods, but that absolute fold change is frequently underestimated by microarray analysis relative to qRT-PCR assay (bold)

Specimen	Gene A (downregulated)		Gene B (upregulated)		Gene C (variable)	
	Microarray	qRT-PCR	Microarray	qRT-PCR	Microarray	qRT-PCR
1	**–3.7**	**–10.9**	1.5	1.4	0.8	–1.8
2	**–3.1**	**–7.4**	2.1	4.7	–1.8	–1.3
3	–2.9	–2.3	**2.3**	**7.4**	2.2	1.6
4	**–5.6**	**–15.9**	2.2	4.4	–1.8	–3.3
6	**–5.2**	**–40.6**	1.3	1.2	**–3.6**	**–52.2**

qualitatively concordant. In our experience, fold-change values identified in microarray experiments are frequently underestimated relative to results obtained by qRT-PCR. That is, a threefold change in gene expression identified by a microarray experiment may appear as a three to twentyfold change in gene expression as determined by qRT-PCR (Table 9.2). As expected, these discrepancies are greatest when comparing microarray data at the very low or very high end of signal intensity. Furthermore, unless qRT-PCR primers are designed to the same target sequence as microarray probes, alternative splicing of cellular transcripts may produce discrepant data.

9.5.6 Troubleshooting

Unless using prevalidated commercial primer/probe assays, the first technical hurdle of the qRT-PCR validation process is to define a robust set of amplification primers for the gene(s) of interest. Problematic primers are those that yield no product (C_T>40 cycles), generate multiple products (a complex melting curve profile), or do not perform efficiently in the range of cDNA template utilized (a nonlinear or attenuated slope of a standard curve plot). Sometimes primers will also generate nonspecific amplification products or (very rarely) specific amplification products with a –RT enzyme negative control. In this latter case, it is important to consider whether some reagent may be contaminated with the target amplicon. It is also possible that the primers may be amplifying trace amounts of genomic DNA that are sometimes present in RNA preparations. When this is identified as the problem, it is generally easier to redesign one or both primers to span a known exon junction rather than trying to treat precious LCM-derived RNA with DNAse. Changing reaction conditions (Mg^{2+} or thermal profile) can increase performance characteristics of some primers, but it is often easier to simply redesign problematic primers. Fortunately, the inexpensive cost and rapid turnaround time of oligonucleotide synthesis greatly facilitate this process. If several rounds of primer synthesis still do not yield a set of primers that amplify their intended target, consider the accuracy of the source sequence and whether there may be some error in annotation from the microarray probe sequence, to the gene annotation of the probe, to the selection of the corresponding reference mRNA sequence, to the accurate sequence and orientation of amplification primers. It also may be useful to design primers to the same target sequence interrogated by the microarray. For example, if a probeset represented on a GeneChip microarray spans a sequence of 170 nucleotides at the 3′ end of a gene transcript, use this 170-nucleotide sequence as the source for qRT-PCR primer design. This strategy should eliminate the possibility of an alternate splicing event causing discrepant results between the two assays. It is also obviously important to perform a set of control reactions with previously validated primers, to ensure that the RNA and serial dilutions of cDNA used for the validation experiments are not degraded.

In the event that validated primers perform poorly with the experimental LCM RNA, it may be instructive to reconfirm the integrity of the RNA and to include appropriate RNA controls, such as diluted RNA isolated from cell lines. We have observed that RNA and especially cDNA stored at low concentrations is not stable after several (more than four) freeze–thaw cycles. Moreover, if primers are to be used with RNA extracted from FFPE tissue, it is important that their performance is validated on FFPE tissue RNA controls.

An inability to technically validate patterns of microarray gene expression using the same RNA in qRT-PCR is not uncommon. As discussed above, quantitative differences in the results between the two methods may occur if signals are not within the linear range of the microarray platform. Qualitative discrepancies may be due to alternative splicing events that are differentially detected by microarray probes and PCR primer sequences or due to errors in gene annotation. However, the most common reason for failure to validate patterns of gene expression identified by microarray analysis is that too few microarray replicates led the investigator to interpret random fluctuations in signals between microarrays as "real" changes in gene expression. In other words, there is no substitute for primary experimental replicates.

Finally, even if patterns of microarray-based gene expression can be validated using the same RNA samples in qRT-PCR assays, this does not confirm the "biological" validity of the result. Since primary microarray analysis may involve an often small and sometimes biased set of specimens, patterns of gene expression identified in a pilot set of dissected cell populations may not be confirmed in a larger set of samples. While not so problematic for genetically defined experimental animal models, this is a formidable problem when dealing with the genetic and environmental variability associated with human clinical specimens. No degree of technical sophistication or troubleshooting can address this issue. This final level of validation can only be achieved by careful study design, higher, more automated throughput LCM of study specimens, and most of all, perseverance.

Acknowledgement. We wish to acknowledge Victoria Holtschlag for her excellent technical assistance in generating the data discussed in this chapter. LCM, tissue processing, microarray hybridization, and quantitative PCR were performed in the Siteman Cancer Center Tissue Procurement and Multiplexed Gene Analysis Core Facilities. This work was supported in part by the Siteman Cancer Center (5P30CA91842-02).

References

Alevizos I, Mahadevappa M, Zhang X, Ohyama H, Kohno Y, Posner M, Gallagher GT, Varvares M, Cohen D, Kim D, Kent R, Donoff RB, Todd R, Yung CM, Warrington JA, Wong DT (2001) Oral cancer in vivo gene expression profiling assisted by laser capture microdissection and microarray analysis. Oncogene 20:6196–6204

Bustin SA (2002) Quantification of mRNA using real-time reverse transcription PCR (RT-PCR): trends and problems. J Mol Endocrinol 29:23–39

Chiang MK, Melton DA (2003) Single-cell transcript analysis of pancreas development. Dev Cell 4:383–393

Coombs NJ, Gough AC, Primrose JN (1999) Optimisation of DNA and RNA extraction from archival formalin-fixed tissue. Nucleic Acids Res 27:e12

Godfrey TE, Kim SH, Chavira M, Ruff DW, Warren RS, Gray JW, Jensen RH (2000) Quantitative mRNA expression analysis from formalin-fixed, paraffin-embedded tissues using 5′ nuclease quantitative reverse transcription-polymerase chain reaction. J Mol Diagn 2:84–91

Hamalainen HK, Tubman JC, Vikman S, Kyrola T, Ylikoski E, Warrington JA, Lahesmaa R (2001) Identification and validation of endogenous reference genes for expression profiling of T helper cell differentiation by quantitative real-time RT-PCR. Anal Biochem 299:63–70

Huang E, Cheng SH, Dressman H, Pittman J, Tsou MH, Horng CF, Bild A, Iversen ES, Liao M, Chen CM, West M, Nevins JR, Huang AT (2003) Gene expression predictors of breast cancer outcomes. Lancet 361:1590–1596

Huang LE, Luzzi V, Ehrig T, Holtschlag V, Watson MA (2002) Optimized tissue processing and staining for laser capture microdissection and nucleic acid retrieval. Methods Enzymol 356:49–62

Iscove NN, Barbara M, Gu M, Gibson M, Modi C, Winegarden N (2002) Representation is faithfully preserved in global cDNA amplified exponentially from sub-picogram quantities of mRNA. Nat Biotechnol 20:940–943

Kabbarah O, Pinto K, Mutch DG, Goodfellow PJ (2003) Expression profiling of mouse endometrial cancers microdissected from ethanol-fixed, paraffin-embedded tissues. Am J Pathol 162:755–762

Karsten SL, Van Deerlin VM, Sabatti C, Gill LH, Geschwind DH (2002) An evaluation of tyramide signal amplification and archived fixed and frozen tissue in microarray gene expression analysis. Nucleic Acids Res 30:E4

Korbler T, Grskovic M, Dominis M, Antica M (2003) A simple method for RNA isolation from formalin-fixed and paraffin-embedded lymphatic tissues. Exp Mol Pathol 74:336–340

Krafft AE, Duncan BW, Bijwaard KE, Taubenberger JK, Lichy JH (1997) Optimization of the isolation and amplification of RNA from formalin-fixed, paraffin-embedded tissue: the Armed Forces Institute of Pathology Experience and Literature Review. Mol Diagn 2:217–230

Leethanakul C, Patel V, Gillespie J, Pallente M, Ensley JF, Koontongkaew S, Liotta LA, Emmert-Buck M, Gutkind JS (2000) Distinct pattern of expression of differentiation and growth-related genes in squamous cell carcinomas of the head and neck revealed by the use of laser capture microdissection and cDNA arrays. Oncogene 19:3220–3224

Livak KJ, Schmittgen TD (2001) Analysis of relative gene expression data using real-time quantitative PCR and the $2^{-\Delta\Delta CT}$ method. Methods 25:402–408

Luzzi V, Holtschlag V, Watson MA (2001) Expression profiling of ductal carcinoma in situ by laser capture microdissection and high-density oligonucleotide arrays. Am J Pathol 158:2005–2010

Luzzi V, Mahadevappa M, Raja R, Warrington JA, Watson MA (2003) Accurate and reproducible gene expression profiles from laser capture microdissection, transcript amplification, and high density oligonucleotide microarray analysis. J Mol Diagn 5:9–14

Ma XJ, Salunga R, Tuggle JT, Gaudet J, Enright E, McQuary P, Payette T, Pistone M, Stecker K, Zhang BM, Zhou YX, Varnholt H, Smith B, Gadd M, Chatfield E, Kessler J, Baer TM, Erlander MG, Sgroi DC (2003) Gene expression profiles of human breast cancer progression. Proc Natl Acad Sci USA 100:5974–5979

Macabeo-Ong M, Ginzinger DG, Dekker N, McMillan A, Regezi JA, Wong DT, Jordan RC (2002) Effect of duration of fixation on quantitative reverse transcription polymerase chain reaction analyses. Mod Pathol 15:979–987

Masuda N, Ohnishi T, Kawamoto S, Monden M, Okubo K (1999) Analysis of chemical modification of RNA from formalin-fixed samples and optimization of molecular biology applications for such samples. Nucleic Acids Res 27:4436–4443

Morrison TB, Weis JJ, Wittwer CT (1998) Quantification of low-copy transcripts by continuous SYBR Green I monitoring during amplification. Biotechniques 24:954–958

Specht K, Richter T, Muller U, Walch A, Werner M, Hofler H (2001) Quantitative gene expression analysis in microdissected archival formalin-fixed and paraffin-embedded tumor tissue. Am J Pathol 158:419–429

Stuart JM, Segal E, Koller D, Kim SK (2003) A gene-coexpression network for global discovery of conserved genetic modules. Science 302:249–255

Vandesompele J, De Preter K, Pattyn F, Poppe B, Van Roy N, De Paepe A, Speleman F (2002) Accurate normalization of real-time quantitative RT-PCR data by geometric averaging of multiple internal control genes. Genome Biol 3:RESEARCH0034

Van Gelder RN, von Zastrow ME, Yool A, Dement WC, Barchas JD, Eberwine JH (1990) Amplified RNA synthesized from limited quantities of heterogeneous cDNA. Proc Natl Acad Sci USA 87:1663–1667

10 Proteomic Applications

David H. Geho, Lance A. Liotta and Emanuel F. Petricoin

10.1 Introduction

The hallmark and distinguishing feature of tissue proteomics is its focus, namely, the direct proteomic study of patient specimens. Rapid technological developments are enabling tissue-based proteomics to characterize distinctive proteomic features present in healthy and diseased clinical materials. These proteomic efforts delve directly into clinical material, seeking patient-based information that could revolutionize clinical care by enabling individualized molecular assessments (Liotta et al. 2001; Petricoin et al. 2002 c).

Primary defects leading to a particular disease, such as a cancer, may reside in the nucleic acid content of a cell. However, the downstream pathophysiology of that disease is determined by alterations in the level of expression or behavior of the protein encoded by a gene. Chromosomal deletions, rearrangements, or mutations are preserved during tumorigenesis because the derangements generate a proteomic environment that is conducive to tumor survival (Hanahan and Weinberg 2000; Hunter 2000; MacDonald et al. 2001). The proteomic environment may elaborate a hyperactive or otherwise defective signaling pathway that enables tumor cells to grow, become motile, and metastasize. As such, diseases often thought of as having significant genetic components, such as cancer, can be alternatively, and fruitfully, recast as proteomic disorders. Consider the aforementioned example. Genetic mutations can cause a cancer cell to grow in an uncontrolled manner through altering the behavior of a key regulatory protein. However, alterations in a particular protein are not restricted to changes in its polypeptide sequence, but may include posttranslational modifications as well. Because of the posttranslational covalent addition of carbohydrates, lipids, or phosphate groups that might aberrantly occur in proteins residing in disease environments, subtle protein-based, disease-related changes would be missed if the evaluation of the disease were limited to a genetic analysis. For instance, an isolated gene array analysis of cancer cells would

Principles and Practice
Emmert-Buck et al. (Eds.):
Dissecting the Molecular Anatomy of Tissue

not reveal the level of important posttranslational modifications, such as phosphorylation, so critical in cell circuitry signaling (Bowden et al. 1999; Hunter 2000; Jeong et al. 2000; Blume-Jensen and Hunter 2001; Celis and Gromov 2003). Even if a gene array detected an increase in the transcription of an enzyme that elaborated certain posttranslational modifications on a protein surface, it would still not be evident which proteins were receiving the modifications and the relative proportion of proteins that were modified.

In tissues, numerous protein–protein interactions mediate communication within a cell's interior and also with its surrounding microenvironment (Pawson 1995; Legrain et al. 2000; Liotta and Petricoin 2000; Schwikowski et al. 2000; Blume-Jensen and Hunter 2001; Ideker et al. 2001; Liotta and Kohn 2001). Protein–protein interactions carry the flow of vital signals that orchestrate cellular behavior. These interactions, in a constant state of fluctuation, allow information to be passed from one protein to the other. The information is contained in modifications that occur during the protein–protein associations. These modifications include phosphorylation and cleavage of peptide bonds, among others. The interactions can be thought of as networks of proteins that form upon certain stimuli and dissociate when the stimuli are withdrawn (Petricoin et al. 2002c). Dysfunctions in the signaling networks can drive a cell towards pathologic behavior as in cancer, enabling the tumor cell to lose growth control, become motile, and metastasize (Liotta and Kohn 2001). A fundamental goal of tissue proteomics is to decipher these circuits. A comprehensive understanding of how protein circuitries behave in disease states will open new doors in revealing pathophysiological mechanisms and in therapeutic design and implementation (Petricoin et al. 2002c).

In health, proteins perform numerous and varied tasks that sustain cellular life. In disease, pathologic protein functions may ultimately destroy the organism. The remarkably diverse roster of cellular protein activities include structural, enzymatic, adhesion, recognition, motility, and signaling functions. The central role proteins play in cellular functions will undoubtedly link specific, unique subsets of the proteome to the pathophysiology of most diseases, either directly or indirectly. Accordingly, understanding how protein populations reflect the state of health of the organism should yield rich clinical dividends, both diagnostically and therapeutically. This type of analysis lies within the realm of tissue proteomics.

The interrelatedness of the tumor with its surrounding microenvironment underscores the importance of studying the proteomic signatures of disease directly in tissue specimens (Liotta and Kohn 2001). In order to most effectively utilize the array of therapeutic agents currently available, it would be optimal to match a drug with its complimentary target from the initiation of therapy rather than undergoing a trial of similar, but distinct agents that may or may not be effective in a particular individual's unique tumor (Karpati et al. 1999; Cheng et al. 2000; Frankel et al. 2000; Gasparini and Gion 2000; Brown and Kirkwood 2001; Cimoli et al. 2001; Kolonin et al. 2001). To achieve this goal, direct evaluation of each patient's clinical biopsy material is required. Historically, a previous impediment to analyzing tissue proteomes has been the limited amounts of patient material available in biopsy specimens. This has been a limitation for proteomic technologies such as electrophoresis, mass spectrometry (MS), and isotope-coded affinity tagging (ICAT) (Gygi et al. 1999; Gorg et al. 2000; Gygi and Aebersold 2000; Hanash 2000). While electrophoresis has been used in limited clinical testing in diseases, such as sickle cell disease, high throughput testing of heterogeneous protein mixtures for diagnostic purposes requires testing platforms that can utilize the scant amounts of tissue present in most biopsies and aspirates. Development of new microtechnologies has been mandated in order to extract the maximal amount of information out of the limited material.

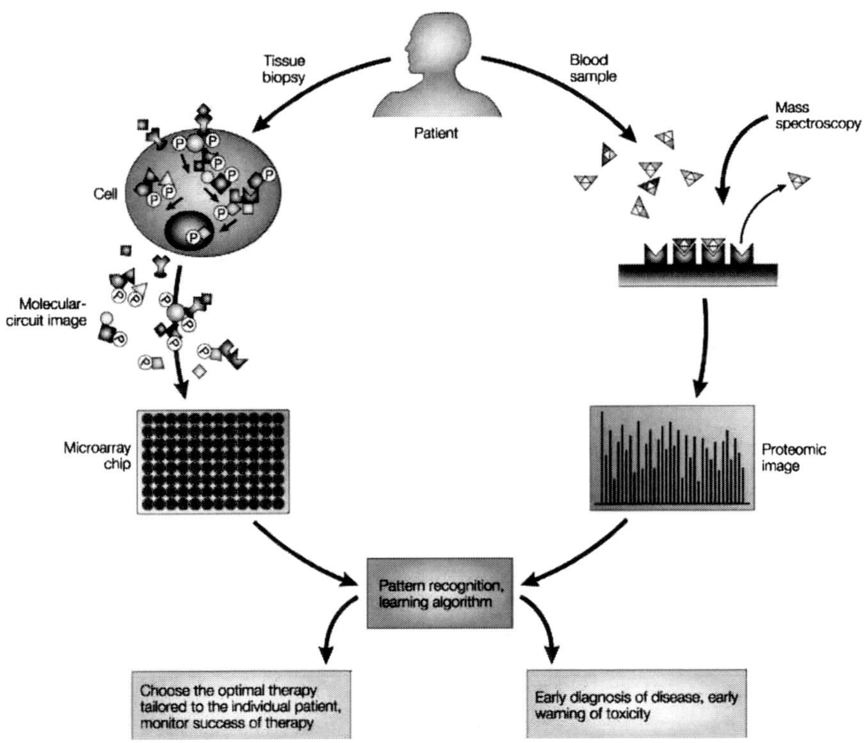

Fig. 10.1. Clinical applications of tissue proteomics. Tissue proteomics tools such as reverse phase protein microarrays and MS pattern diagnostics can be applied to diagnosis and therapeutic assessments. Biopsied materials or body fluid specimens can be profiled using these complimentary approaches, thereby enabling personalized, combinatorial therapies to be employed (see color illustration at the end of the book) (Figure from Petricoin et al. 2002c)

In the young discipline of tissue proteomics, two major issues are the objects of current study. First, can direct proteomic examination of clinical specimens be reproducibly performed in patient-care settings? Secondly, in proteomic pattern diagnosis, do the signatures of diseased tissues effectively predict clinical outcomes? These questions are currently being assessed using two technological partners, namely, protein microarrays (Ge 2000; MacBeath and Schreiber 2000; Paweletz et al. 2001; Zhu and Snyder 2001, 2003; Charboneau 2002; Lal et al. 2002; MacBeath 2002; Cutler 2003; Espina et al. 2003; Liotta et al. 2003 a; Wilson and Nock 2003) and mass spectrometry (MS) proteomic profiling (Adam et al. 2002; Chaurand and Caprioli 2002; Li et al. 2002; Petricoin et al. 2002 a, b; Conrads et al. 2003; Hingorani et al. 2003; Petricoin and Liotta 2003; Yanagisawa et al. 2003; Schwartz et al. 2004; Fig. 10.1, see also color illustration at the end of the book). These technologies are well-paired, for they explore these questions from two different perspectives. With protein microarrays, well-defined protein effectors of important cell signaling activity are surveyed for levels of activity in well-defined disease states. On the other hand, MS pattern proteomics starts with no preliminary understanding of the identity of proteins that are surveyed. Coupled with bioinformatics computer programs, MS patterns yield important diagnostic information in varying diseases and tissue types. Rapid technical advances in diverse methodologies are driving the field of proteomics closer and closer to clinical applicability, the underlying goal of tissue proteomics. These technologies will help translate proteomics into bedside applicability through defining drug targets, assessing response to treatments, and early diagnostics (Liotta et al. 2001).

10.2 Laser Capture Microdissection

A major contribution to the field of tissue proteomics was the development of laser capture microdissection (Emmert-Buck et al. 1996). Sites of disease within a tissue have polymorphic populations of cells, with some being diseased cells and others surrounding stromal cells. The surrounding cells can be epithelial, endothelial, hematopoietic, and mesenchymal in derivation. Each type of cell contributes its own unique palette of proteins into the surrounding interstitium. In order to parse the molecular events occurring within such a complex environment, it is essential to be able to purify distinct subsets of cellular populations for study. Studies of tumor cells grown in culture and microdissected tumor cells have demonstrated that there are marked molecular differences generated by the two types of growth conditions (Ornstein et al. 2000). LCM technology has been used to define numerous new molecules that can serve as drug targets (Emmert-Buck et al. 1996, 2000 a; Ornstein et al. 2000; Knezevic et al. 2001; Wulfkuhle et al. 2001, 2002, 2003 b; Ahram et al. 2002; Craven et al. 2002; Jones et al. 2002). LCM is an essential component in tissue proteomics and has provided pure patient-derived specimens for both protein microarrays and MS-based protein pattern diagnostics.

10.3 Protein Microarrays

Protein microarrays provide a high throughput means of measuring many different tissue-derived proteins on the same surface, or microarray. Two formats for protein microarrays have been developed (Ge 2000; Liotta and Petricoin 2000; MacBeath and Schreiber 2000; Paweletz et al. 2001; Zhu and Snyder 2001, 2003; Charboneau 2002; Lal et al. 2002; MacBeath 2002; Cutler 2003; Wilson and Nock 2003; Fig. 10.2, see also color illustration at end of book). One of them, the forward phase array, has a bait molecule

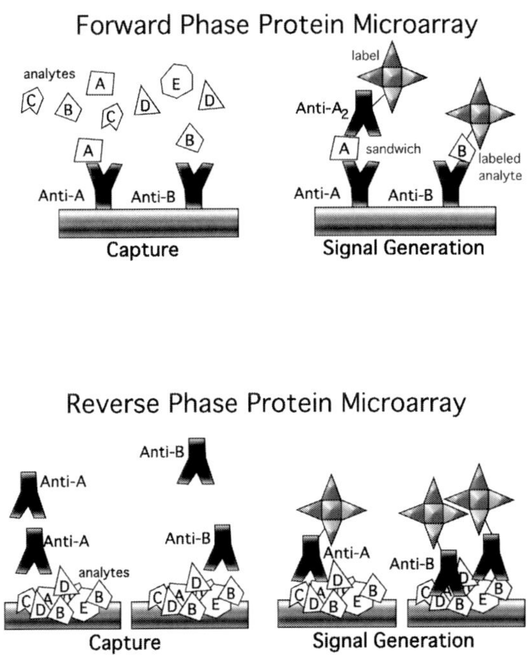

Fig. 10.2. Protein microarrays. With forward phase arrays, a capture, or bait molecule is immobilized on a substrate. Analytes in a solution are then incubated on the array. Using another antibody, binding of analytes is determined. Alternatively, the analyte itself may be labeled. An alternative approach is the reverse-phase array. In this platform, analytes are immobilized directly on a substrate. Then a probed molecule, such as an antibody, is incubated on the array. Following this step, the presence of the probe is detected by a labeled secondary molecule such as an antibody. For low abundance molecules, the signal can be amplified (see color illustration at the end of the book) (Figure from Liotta et al. 2003 a)

attached to the array surface (Liotta et al. 2003a). The surface is then incubated with a mixture of analytes. Analytes that bind are then detected with a labeled probe. This approach requires that two different binding molecules, such as antibodies, be used in a sandwich-type assay. Finding two antibodies with distinct epitopes on the same target analyte can be difficult, given the limited spectrum of commercially available high quality antibodies.

The alternative approach is the reverse phase array (RPA; Paweletz et al. 2001; Charboneau 2002; Grubb et al. 2003; Wulfkuhle et al. 2003a). In this type of array format, multiple analytes, such as extracted cellular lysate components, are immobilized onto an array surface directly. Each spot is its own self-contained specimen, allowing multiple test spots to be probed by an individual detection molecule on the same array. In tissue-based proteomics, systems that maximize the use of limited patient material are selected over systems that have higher sample requirements. In this setting, the reverse phase array has emerged as a primary vehicle for measuring activity levels of protein signaling pathways in tumor cells.

Reverse phase microarray technologies, though in constant state of improvement and development, have been reduced to basic, well-practiced methodologies. These microarrays consist of a substrate, such as nitrocellulose, silica, or nylon (Stillman and Tonkinson 2000; Tonkinson and Stillman 2002; Espina et al. 2003), upon which proteins are applied in discrete areas, or spots (Espina et al. 2003). The qualities of the surface of the array should include low background and a large surface area to maximize binding of protein. Moreover, the surface chemistry should not significantly alter the conformational structure of the bait molecules (Espina et al. 2003). As a substrate, nitrocellulose offers a high surface area, low background, and does not significantly interfere with detection of the bait molecule (Espina et al. 2003). Nitrocellulose binds very tightly to proteins through electrostatic interactions. Because of this, it is difficult to strip the surface for multiple uses (Stillman and Tonkinson 2000; Tonkinson and Stillman 2002). Using arrayers such as the Affymatrix (GMS417, Affymetrix/MWG Biotech), many replicate slides can be printed of the same clinical material. Depending on the application, the spot may consist of a mixture of molecules or a homogenous population. For clinical purposes, the spotted material is most often a heterogeneous mixture of proteins derived from cellular lysates harvested using laser capture microdissection.

For example, a core needle biopsy (16–18 gauge needle) from a cancer patient is flash frozen and stored at -80 °C for preservation purposes. These biopsies yield roughly 2000–8000 cells (Espina et al. 2003). As required, proteases can be delivered to the staining elements or in extraction buffers (Simone et al. 2000). Specimens are frozen rather than fixed in formalin because fixation can drastically alter the properties of proteins that are studied using the microarray technique (Emmert-Buck et al. 2000b; Gillespie et al. 2002; Ahram et al. 2003). The frozen material is then sectioned at 5–10 μm thicknesses, yielding frozen sections that can be stained. Laser capture microdissection using a machine such as the AutoPix (Arcturus Engineering, Inc., Mountain View, CA) is then performed on the frozen section material from the original biopsy that has been completely dehydrated using alcohol and xylene (Emmert-Buck et al. 1996, 2000b). This allows pure populations of tumor cells or stromal cells to be procured and studied. It is estimated that approximately 5000 cells/10 μl of lysis buffer are required for successful array production (Espina et al. 2003). Stringent extraction of proteins from lysed microdissected cells is required to liberate as many proteins as is possible (see review, Espina et al. 2003).

Using an arrayer, spots comprising a minute specimen of a cellular proteome are placed on the substrate surface. Spotting is often done such that a dilution curve of lysate sample is present in triplicate on the substrate. Because many spots can be arrayed on the same slide, it is possible to compare a patient's material before and after treatment with a therapeutic agent or to compare groups of patients receiving the same treatment. For quality control purposes, a recombinant protein, a peptide, or a reference lysate can be spotted on the array. Such a reference allows comparisons to be made across test samples for each of the probing antibodies. A new format for protein microarrays, the sector array, will significantly expand the amount of information that can be gleaned from an individual microarray slide (Espina et al. 2003). This format entails placement of a reservoir around a substrate pad. A different antibody can be used to probe each of the sectors. Since multiple substrate areas are present on a single array surface, multiplexed probing with different antibodies can occur on the same array surface.

After spotting, the array surface can then be probed with molecules, such as antibodies that specifically recognize proteins of interest within the lysate. Each slide is probed with a separate antibody, since nitrocellulose is not amenable to stripping and reprobing. An important aspect of microarray implementation is acquisition of specific ligands (antibodies) that bind with high affinities (Templin et al. 2002). An initial step in assessing the applicability of a particular antibody in a microarray setting is to validate its specificity and relative affinity using a Western blot assay.

Multiple detection systems can be employed, including chromogenic, fluorometric, radioisotopic, and luminescent systems (Espina et al. 2003; Kukar et al. 2002; Liotta et al. 2003a; Morozov et al. 2002; Wiese 2003). Because proteins of interest are often in low abundance, amplification strategies have been utilized in the detection arm of the assay (King et al. 1997; Schweitzer et al. 2002). One such detection strategy is the catalyzed reporter deposition technology (Bobrow et al. 1989, 1991; Hunyady et al. 1996; King et al. 1997; Graf and Friedl 1999). The presence of the probe provides a signal, or a positive spot. A comparison of the degree of positive signals is made across the array. The intensity of the probe signal is proportional to the number of labeled probes bound to the spot's bait molecule (Espina et al. 2003).

For data analysis, the array is scanned, the signal intensity is assessed, and a normalized data value is assigned for each spot. This enables comparison of signal intensity across an array. Total protein present in a spot is assessed using a SyproRuby blot assay (Molecular Probes, Eugene, OR). This allows adjustment for variability in protein amounts in various samples harvested by LCM. Data generated from the microarray can then be studied using the PSCAN system (Carlisle et al. 2000; Liotta et al. 2003 a).

All new technologies face challenges, and protein microarrays are no exception to this rule. For one, the proteins to be studied can vary in intracellular concentrations by up to a factor of 10^{10}. Therefore, it is important to develop probes, such as antibodies that will detect low abundance proteins with high specificity and sensitivity. It has been estimated that sensitivity within the femtomolar range is required for some signaling proteins (Schweitzer et al. 2002). Moreover, antibodies must be developed that effectively demonstrate subtle posttranslational modifications. The Human Proteome Organization (HUPO) is leading an effort to produce and characterize antibody libraries to meet these needs (Hanash 2003; Tyers and Mann 2003). Further, cellular lysates contain molecules that can confound protein microarray detection systems, such as peroxidases, avidin, biotin, immunoglobulins, fluorescent molecules, and alkaline phosphatases (Espina et al. 2003).

Reverse phase arrays have a number of advantages over other array types. For one, denatured lysates can be used in RPAs, which is not the case for tissue arrays. On the other hand, nondenatured lysates can be used in RPAs as well, which allow macromolecular complexes of diverse biomolecular complexes that do not survive denaturants to be studied. Another benefit of the RPA is that dilution curves are built into the spotting of the cellular lysates. This allows the linear component of the antibody–analyte binding event to be quantitatively ascertained. Finally, since the cellular lysates in the reverse phase arrays are not labeled, the proteins of interest are not modified, helping to generate a more reproducible assay (Petricoin et al. 2002 c; Espina et al. 2003).

Presently, reverse phase microarrays are now being integrated into clinical trials at the National Cancer Institute. Reverse phase microarrays enable profiling of posttranslational modifications of proteins involved in signaling pathways. For example, as mentioned previously, phosphorylation of proteins is a primary means for cellular signals to be transmitted. Activation of signaling pathways can be assessed by direct comparisons of the phosphorylated, or active, form of a signaling protein versus the total amount of the same protein. Thus, by studying the activation state of important proteins, or "nodes", within a signaling network, trends can be observed in disease states as well as after treatments (Liotta et al. 2001; Petricoin et al. 2002 c). As more and more pathways are described and correlated with disease states, physicians will be able to employ their understanding of the molecular anatomy of an organism. Familiarity with the signaling infrastructure of a patient's tumor cells will enable individualized therapeutic plans to be developed and will serve as a springboard for drug discovery (Liotta et al. 2001; Petricoin et al. 2002 c).

For example, reverse phase arrays have been applied to investigating fundamental aspects of tumor pathophysiology. One previously addressed question was whether premalignant transformation of cells is due to decreased apoptosis (via activation of apoptosis-inhibiting signaling pathways), or due to activation of mitogenic growth pathways (Paweletz et al. 2001). In this study using prostate cancer specimens, LCM was used to isolate cells from patient-matched normal epithelium, premalignant epithelium, and invasive prostate cancer. Microarray analysis of the specimens demonstrated that phosphorylation, and thus activation, of the signaling protein Akt is a pivotal event in the development of prostate cancer. Thus, on histopathologic slides the cell accumulations within prostate glands, or prostatic intraepithelial neoplasia, is not caused by activation of mitogenic pathways, but rather by activation of apoptosis-inhibiting pathways. This study demonstrates how reverse phase protein microarrays can be used to drive drug discovery efforts. Drugs that target Akt function might have therapeutic applicability in treating prostatic disease.

Proteomic techniques have recently been brought to bear on other interesting, unresolved areas of study. For example, tissue proteomic work has been performed on the classical question of tumor-altered metabolism. For the first time since Warburg's initial work in the 1920s implicating a greater reliance of tumor cells on anaerobic function (Warburg et al. 1924), direct comparison of some of the metabolic machinery in tumor and normal cells has been made (Herrmann et al. 2003). Without the tissue proteomic techniques of laser capture microdissection (LCM) and reverse phase protein microarrays, such studies would be almost impossible. Another interesting area of research facilitated by advances in tissue proteomics is an assessment of the proteomic content of lymph (Leak et al. 2002).

10.4 Mass Spectrometry in Tissue Proteomics

Mass spectrometry analysis of human tissues is a rapidly growing field within tissue proteomics. MS has applicability both for biomarker discovery and development of clinical diagnostics. A major focus within the field of tissue proteomics is MS protein profiling (Paweletz et al. 2000; Stoeckli et al. 2001; Chaurand and Caprioli 2002; Petricoin et al. 2002 a; Yanagisawa et al. 2003; Schwartz et al. 2004). In MS pattern-based diagnostics, the pattern within a mass spectra holds the diagnostic key, even without knowledge of the identity of molecules generating the discriminating pattern. This approach provides the potential for high throughput assessments using small amounts of patient material and has been used to classify tissues into groups that match previously derived histopathological diagnoses.

MS protein pattern diagnostics have been applied to body tissues, both solid tissue specimens and body fluids (Adam et al. 2002; Li et al. 2002; Petricoin et al. 2002b, c; Conrads et al. 2003; Hingorani et al. 2003; Petricoin and Liotta 2003).

MS technologies include surface-enhanced laser desorption/ionization (SELDI; Fig. 10.3, see also color illustration at end of book). With this assay system a subset of proteins within a heterogeneous mixture binds to a chip surface (which has chemically based retention properties). The proteins may be from a cell lysate or a body fluid specimen. Unbound proteins are removed by washing. Proteins bound to the chip are then covered with a matrix followed by drying and crystallization. The chip is then placed into a vacuum chamber. A laser beam directed at the chip and the matrix transfers energy to the protein, ionizing and desorbing it (Hillenkamp and Karas 1990). Once an ion is launched, the time of flight required for the ion to reach a detection element is measured generating a mass to charge value with smaller ions reaching the detection element earlier than the larger species.

An initial MS pattern study was performed on microdissected human solid tumor cells (Paweletz et al. 2000). Laser capture microdissection was used to isolate pure tumor cells. The proteins from the cells were extracted and the lysate was dried upon a prepared surface for surface-enhanced laser desorption ionization time-of-flight (SELDI-TOF) spectrometry. This work used cellular lysates from a limited number of microdissected cells to generate MS patterns that, when interrogated by bioinformatic programs, distinguished early from late malignancy. MS has also been directly applied to tissue slices, which obviates the necessity for LCM (Stoeckli et al. 2001; Chaurand and Caprioli 2002; Yanagisawa et al. 2003; Schwartz et al. 2004). In this technique, frozen sections of clinical specimens are dried onto a matrix assisted laser desorption/ionization (MALDI) plate. The specimen is then coated with an ionization matrix. After this preparation, a laser is directly applied to the specimen in a vacuum chamber. Thus, direct probing of the tissue specimen results in the mass spectral pattern. This approach has been used for lung and brain tumors (Yanagisawa et al. 2003; Schwartz et al. 2004), both of which had MS patterns that bioinformatic programs placed into tumor and normal groupings, which corresponded to previous histologic diagnoses.

Another use of MS-based proteome diagnostics is disease detection through assessment of body fluids, such as blood components. The approach was first described for the serum-based diagnostic classification of ovarian cancer (Petricoin et al. 2002b). Serum samples were subjected to MS analysis followed by bioinformatic analysis. This approach was able to differentiate patients with cancer from those who had no malignancy. This technique was even able to distinguish those patients with early disease stage. This approach has been applied to numerous other disease states as well (Adam et al. 2002; Li et al. 2002; Petricoin et al. 2002a; Con-

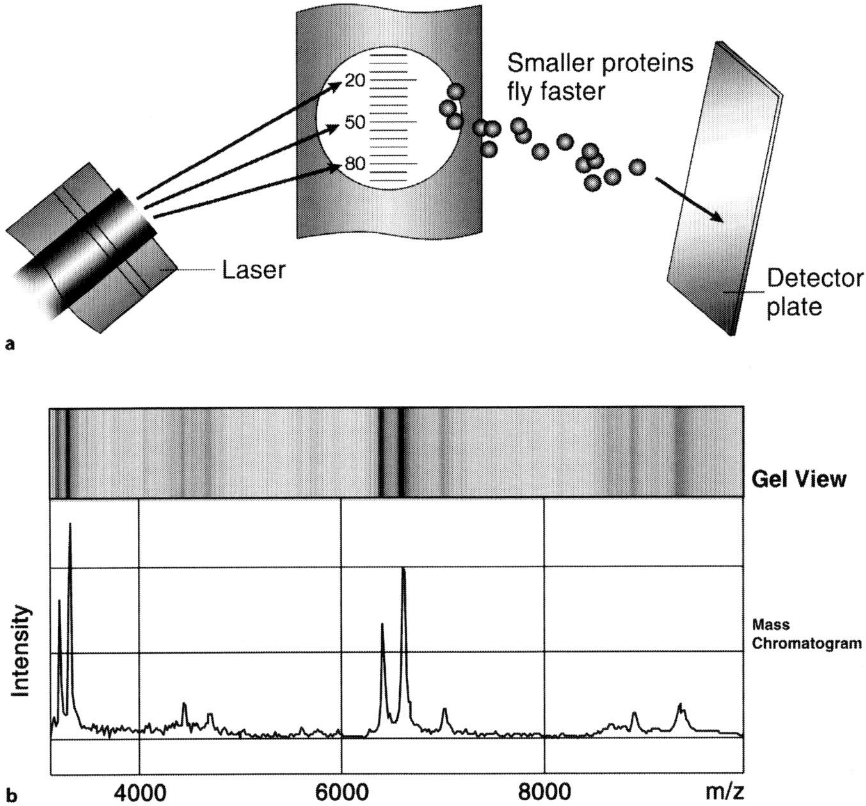

Fig. 10.3 a, b. Mass spectroscopy using a SELDI-TOF format. For the sake of reproducibility and high throughput testing, a robotic system delivers 1 μl of native serum to a protein-binding chip surface. The surface chemistry of the chip immobilizes a subset of the proteins from the heterogeneous mixture (**a**). Unbound proteins are removed by washing and a matrix is applied to the remaining, bound proteins (**a**). After washing and drying, the chip is irradiated with a laser in a vacuum chamber. The proteins, desorbed by the laser, are launched in a charged ionic state. The length of time required for the ion to reach a detector provides a measure of the mass to charge (m/z) value. A proteomic signature of the serum emerges comprised of up to 15,500 data points (data readings between 500 and 20,000 m/a; **b**). Higher resolution MS technologies generate many more data points (up to 400,000 data points for an m/z range of 500–12,000). Using bioinformatics techniques, the data can be analyzed for the presence of diagnostic serum patterns (see color illustration at the end of the book) (Figure from Wulfkuhle et al. 2003)

rads et al. 2003; Hingorani et al. 2003; Petricoin and Liotta 2003). Higher resolution MS technology has recently become available that is refining and enhancing the MS approach. These MS machines utilize hybrid quadruple time of flight mass spectrometer (QSTAR pulsar *i*, Applied Biosystems Inc., Framingham, MA) fitted with a ProteinChip array interface (Ciphergen Biosystems Inc.). Systems such as these provide much richer resolution, which can generate spectral patterns yielding increased sensitivity and specificity when applied to ovarian cancer detection (Conrads et al. 2003). Even low resolution MS proteomic profiles generate massive data sets. These require data mining by bioinformatic programs that interrogate the data and employ pattern recognition systems (Petricoin et al. 2002b).

The proteome expressed in a disease state reflects disease-specific alterations. As the proteome is turned over through enzymatic degradation, a subset of degradation products likely seeps into the interstitium and then into the circulation. As such, the serum represents a rich source of low molecular weight biomarkers that reflect organismal processes (Liotta et al. 2003b; Mehta et al. 2003). Recent work indicates that these low molecular weight biomarker molecules become associated with carrier proteins, such as albumin, which allow them to survive renal clearance mechanisms in the serum (Liotta et al. 2003b; Mehta et al. 2003). As such, numerous intact and degraded proteins comprise the serum proteome. There is reinvigorated interest in characterizing the serum proteome (Tirumalai et al. 2003). In the past, biomarker discovery focused on trying to find single markers that distinguished normal from disease (Adam et al. 2001; Xiao et al. 2001; Anderson and Anderson 2002; Carter et al. 2002; Kim et al. 2002; Rosty et al. 2002). It is most likely that proteins within a cancer patient are modified forms of normal proteins. As such, defining a single marker for a given cancer may not be as fruitful an approach as seeking patterns of protein modifications that reflect the widespread effects of disease within a patient. Because of this, abundant serum proteins, previously thought of as contaminants, are now being sought for the cargo they carry, such as degraded forms of proteins carrying disease-specific alterations.

In the future, custom-designed nanoparticles will be developed that mimic the low molecular weight molecule binding properties of native serum carrier proteins (Liotta et al. 2003b). These will be introduced into serum to seek out and bind low molecular weight signatures of disease, effectively amplifying them for MS-based analysis. These particles could directly interface with MS platforms and be joined with new ultra-high resolution MS such as Fourier transform ion cyclotron resonance MS (Shen et al. 2001, 2004). This approach would allow for both the diagnostic analysis as well as identification of the protein species contributing to diagnostic discrimination.

Integration of MS into patient care settings will require thorough evaluation of the reproducibility of the technology in diverse environments. Because body fluids are labile, uniform handling will be required for diagnostic specimens as MS technology is able to detect such subtle distinctions in structures. Reference standards must be developed for calibration of systems. For these purposes, the NCI-FDA Clinical Proteomics program uses a pooled reference standard from NIST (SRM-1951A) applied at random on each array.

MS pattern-based proteomics is a high throughput technique with direct applicability to clinical settings. For one, it could be possible to process a vast number of samples on each MS instrument. This can be achieved even with limited patient materials, such as a blood sample. The patterns generated by MS proteomics yield multiple potential biomarkers that provide an increased level of discrimination when compared to a single biomarker, which is important when dealing with large patient groups. As MS technology develops further, it will be possible to not only generate diagnostic patterns, but also gain identity data on particular MS ion spectra. This will usher in a new era of biomarker and therapeutic target discovery. The hallmarks of MS diagnostics are early detection using limited patient material, elements that are highly desirable in clinical settings.

10.5 Conclusions

Tissue proteomics utilizes complimentary technical tools to decipher the many roles that proteins play in diseases. This field has much broader applicability than solely creating a list of proteins found in various pathological conditions. Rather, it has the opportunity of being directly applied within clinical decision-making settings. Early diagnosis of disease through evaluation of serum proteomic disease signatures could radically transform the stage at which many diseases are detected. For example, ovarian cancer is more often than not discovered when the disease has reached an advanced stage (Hoskins et al. 2000). At that point, the cancer cells have metastasized and 5-year survival rates are dismal. On the other hand, early stage disease can be treated effectively (Bast et al. 1983; Jacobs et al. 1999; Menon and Jacobs 2000; Cohen et al. 2001; Wulfkuhle et al. 2003b). Obviously, early, preclinical detection of ovarian cancer via serum-based MS pattern diagnostics is an exciting alternative to diagnosis when the disease is clinically apparent (Petricoin et al. 2002b).

Protein microarrays are paving the way for proteomic assessments of biopsy materials in cancer patients. The goal of this work is to thoroughly define the numerous signaling pathways that run in and out of a cell and correlate changes in the pathways with disease entities. Understanding the molecular anatomy of this signaling infrastructure will yield drug targets, biomarker candidates, and prognostic information. As multiple important signaling proteins, or nodes, within a pathway are delineated, the possibility of treating multiple nodes within a deranged signaling pathway using a combinatorial therapy approach may yield higher efficacy and fewer side effects (Liotta and Petricoin 2000; Schwikowski et al. 2000; Ideker et al. 2001).

The most relevant uses of tissue proteomic techniques are focused upon patient care. A good physician understands that each patient is unique and requires specialized care. One important aspect of each patient's uniqueness has been locked away until now, namely, each patient's molecular individuality. Tissue proteomics is unlocking this realm. Soon, it is hoped, doctors will have access to the molecular anatomy of their patient's disorder, just as they now have access to tissue-based histopathologic assessments. In the future, molecular anatomy profiles will enable and guide individualized patient assessments and therapeutic interventions, a process enabled by present research in tissue proteomics.

References

Adam BL, Vlahou A, Semmes OJ et al. (2001) Proteomic approaches to biomarker discovery in prostate and bladder cancers. Proteomics 1:1264–1270

Adam BL, Qu Y, Davis JW et al. (2002) Serum protein fingerprinting coupled with a pattern-matching algorithm distinguishes prostate cancer from benign prostate hyperplasia and healthy men. Cancer Res 62:3609–3614

Ahram M, Best CJ, Flaig MJ et al. (2002) Proteomic analysis of human prostate cancer. Mol Carcinog 33:9–15

Ahram M, Flaig MJ, Gillespie JW et al. (2003) Evaluation of ethanol-fixed, paraffin-embedded tissues for proteomic applications. Proteomics 3:413–421

Anderson NL, Anderson NG (2002) The human plasma proteome: history, character, and diagnostic prospects. Mol Cell Proteomics 1:845–867

Bast RC Jr, Klug TL, St John E et al. (1983) A radioimmunoassay using a monoclonal antibody to monitor the course of epithelial ovarian cancer. N Engl J Med 309:883–887

Blume-Jensen P, Hunter T (2001) Oncogenic kinase signalling. Nature 411:355–365

Bobrow MN, Harris TD, Shaughnessy KJ et al. (1989) Catalyzed reporter deposition, a novel method of signal amplification. Application to immunoassays. J Immunol Methods 125:279–285

Bobrow MN, Shaughnessy KJ, Litt GJ et al. (1991) Catalyzed reporter deposition, a novel method of signal amplification. II. Application to membrane immunoassays. J Immunol Methods 137:103–112

Bowden ET, Barth M, Thomas D et al. (1999) An invasion-related complex of cortactin, paxillin and PKCmu associates with invadopodia at sites of extracellular matrix degradation. Oncogene 18:4440–4449

Brown CK, Kirkwood JM (2001) Targeted therapy for malignant melanoma. Curr Oncol Rep 3:344–352

Carlisle AJ, Prabhu VV, Elkahloun A et al. (2000) Development of a prostate cDNA microarray and statistical gene expression analysis package. Mol Carcinog 28:12–22

Carter D, Douglass JF, Cornellison CD et al. (2002) Purification and characterization of the mammaglobin/lipophilin B complex, a promising diagnostic marker for breast cancer. Biochemistry 41:6714–6722

Celis JE, Gromov P (2003) Proteomics in translational cancer research: toward an integrated approach. Cancer Cell 3:9–15

Charboneau L (2002) Utility of reverse phase protein microarrays: applications to signaling pathways and human body arrays. Brief Funct Genomics Proteomics 1:305–315

Chaurand P, Caprioli RM (2002) Direct profiling and imaging of peptides and proteins from mammalian cells and tissue sections by mass spectrometry. Electrophoresis 23:3125–3135
Cheng JD, Rieger PT, von Mehren M et al. (2000) Recent advances in immunotherapy and monoclonal antibody treatment of cancer. Semin Oncol Nurs 16[4 Suppl 1]:2–12
Cimoli G, Bagnasco L, Pescarolo MP et al. (2001) Signaling proteins as innovative targets for antineoplastic therapy: our experience with the signaling protein c-myc. Tumori 87:S20–S23
Cohen LS, Escobar PF, Sharm C et al. (2001) Three-dimensional power Doppler ultrasound improves the diagnostic accuracy for ovarian cancer prediction. Gynecol Oncol 821:40–48
Conrads TP, Zhou M, Petricoin EF et al. (2003) Cancer diagnosis using proteomic patterns. Expert Rev Mol Diagn 3:411–420
Craven RA, Totty N, Harnden P et al. (2002) Laser capture microdissection and two-dimensional polyacrylamide gel electrophoresis: evaluation of tissue preparation and sample limitations. Am J Pathol 160:815–822
Cutler P (2003) Protein arrays: the current state-of-the-art. Proteomics 3:3–18
Emmert-Buck MR, Bonner RF, Smith PD et al. (1996) Laser capture microdissection. Science 274:998–1001
Emmert-Buck MR, Gillespie JW, Paweletz CP et al. (2000) An approach to proteomic analysis of human tumors. Mol Carcinog 27:158–165
Emmert-Buck MR, Strausberg RL, Krizman DB et al. (2000) Molecular profiling of clinical tissues specimens: feasibility and applications. J Mol Diagn 2:60–66
Espina V, Mehta AI, Winters ME et al. (2003) Protein microarrays: molecular profiling technologies for clinical specimens. Proteomics 3:2091–2100
Frankel AE, Sievers EL, Scheinberg DA et al. (2000) Cell surface receptor-targeted therapy of acute myeloid leukemia: a review. Cancer Biother Radiopharm 15:459–476
Gasparini G, Gion M (2000) Molecular-targeted anticancer therapy: challenges related to study design and choice of proper endpoints. Cancer J 6:117–131
Ge H (2000) UPA, a universal protein array system for quantitative detection of protein–protein, protein–DNA, protein–RNA and protein–ligand interactions. Nucleic Acids Res 28:e3
Gillespie JW, Best CJ, Bichsel VE et al. (2002) Evaluation of non-formalin tissue fixation for molecular profiling studies. Am J Pathol 160:449–457
Gorg A, Obermaier C, Boguth G et al. (2000) The current state of two-dimensional electrophoresis with immobilized pH gradients. Electrophoresis 21:1037–1053
Graf R, Friedl P (1999) Detection of immobilized proteins on nitrocellulose membranes using a biotinylation-dependent system. Anal Biochem 273:291–297
Grubb RL, Calvert VS, Wulkuhle JD et al. (2003) Signal pathway profiling of prostate cancer using reverse phase protein arrays. Proteomics 3:2142–2146
Gygi SP, Aebersold R (2000) Mass spectrometry and proteomics. Curr Opin Chem Biol 4:489–494
Gygi SP, Rist B, Gerber SA et al. (1999) Quantitative analysis of complex protein mixtures using isotope-coded affinity tags. Nat Biotechnol 17:994–999
Hanahan D, Weinberg RA (2000) The hallmarks of cancer. Cell 100:57–70
Hanash S (2003) Disease proteomics. Nature 422:226–232
Hanash SM (2000) Biomedical applications of two-dimensional electrophoresis using immobilized pH gradients: current status. Electrophoresis 21:1202–1209

Herrmann PC, Gillespie JW, Charboneau L et al. (2003) Mitochondrial proteome: altered cytochrome c oxidase subunit levels in prostate cancer. Proteomics 3(9):1801–1810

Hillenkamp F, Karas M (1990) Mass spectrometry of peptides and proteins by matrix-assisted ultraviolet laser desorption/ionization. Methods Enzymol 193:280–295

Hingorani SR, Petricoin EF, Maitra A et al. (2003) Preinvasive and invasive ductal pancreatic cancer and its early detection in the mouse. Cancer Cell 4:437–450

Hoskins WJ, Perez CA et al. (eds) (2000) Tumor markers. Lippincott, Williams and Wilkins, Philadelphia

Hunter T (2000) Signaling - 2000 and beyond. Cell 100:113–127

Hunyady B, Krempels K, Maitra A et al. (1996) Immunohistochemical signal amplification by catalyzed reporter deposition and its application in double immunostaining. J Histochem Cytochem 44:1353–1362

Ideker T, Thorsson V, Ranish JA et al. (2001) Integrated genomic and proteomic analyses of a systematically perturbed metabolic network. Science 292:929–934

Jacobs IJ, Skates SJ, MacDonald N et al. (1999) Screening for ovarian cancer: a pilot randomised controlled trial. Lancet 353:1207–1210

Jeong H, Tombor B, Albert R et al. (2000) The large-scale organization of metabolic networks. Nature 407:651–654

Jones MB, Krutzsch H, Shu H et al. (2002) Proteomic analysis and identification of new biomarkers and therapeutic targets for invasive ovarian cancer. Proteomics 2:76–84

Karpati G, Li H, Nalbantoglu J et al. (1999) Molecular therapy for glioblastoma. Curr Opin Mol Ther 1:545–552

Kim JH, Skates SJ, Uede T et al. (2002) Osteopontin as a potential diagnostic biomarker for ovarian cancer. JAMA 287(13):1671–1679

King G, Payne S, Walker F et al. (1997) A highly sensitive detection method for immunohistochemistry using biotinylated tyramine. J Pathol 183(2):237–41

Knezevic V, Leethanakul C, Bichsel VE et al. (2001) Proteomic profiling of the cancer microenvironment by antibody arrays. Proteomics 1:1271–1278

Kolonin M, Pasqualini R, Arap W et al. (2001) Molecular addresses in blood vessels as targets for therapy. Curr Opin Chem Biol 5:308–313

Kukar T, Eckenrode S, Gu Y et al. (2002) Protein microarrays to detect protein-protein interactions using red and green fluorescent proteins. Anal Biochem 306:50–54

Lal SP, Christopherson RI, dos Remedios CG et al. (2002) Antibody arrays: an embryonic but rapidly growing technology. Drug Discov Today 7[18 Suppl]:S143–S149

Leak LV, Petricoin EF 3rd, Jones M et al. (2002) Proteomic technologies to study diseases of the lymphatic vascular system. Ann N Y Acad Sci 979:211–228; discussion 229–234

Legrain P, Jestin JL, Schachter V et al. (2000) From the analysis of protein complexes to proteome-wide linkage maps. Curr Opin Biotechnol 11:402–407

Li J, Zhang Z, Rosenzweig J et al. (2002) Proteomics and bioinformatics approaches for identification of serum biomarkers to detect breast cancer. Clin Chem 48:1296–1304

Liotta L, Petricoin E (2000) Molecular profiling of human cancer. Nat Rev Genet 1:48–56

Liotta LA, Kohn EC (2001) The microenvironment of the tumour-host interface. Nature 411:375–379

Liotta LA, Kohn EC, Petricoin EF et al. (2001) Clinical proteomics: personalized molecular medicine. JAMA 286:2211–2214
Liotta LA, Espina V, Mehta AI et al. (2003) Protein microarrays: Meeting analytical challenges for clinical applications. Cancer Cell 3:317–325
Liotta LA, Ferrari M, Petricoin E et al. (2003) Clinical proteomics: written in blood. Nature 425:905
MacBeath G (2002) Protein microarrays and proteomics. Nat Genet 32[Suppl]:526–532
MacBeath G, Schreiber SL (2000) Printing proteins as microarrays for high-throughput function determination. Science 289:1760–1763
MacDonald TJ, Brown KM, Petricoin E et al. (2001) Expression profiling of medulloblastoma: PDGFRA and the RAS/MAPK pathway as therapeutic targets for metastatic disease. Nat Genet 29:143–152
Mehta AI, Ross S, Lowenthal M et al. (2003) Biomarker amplification by serum carrier protein binding. Disease Markers 19:1–10
Menon U, Jacobs IJ (2000) Recent developments in ovarian cancer screening. Curr Opin Obstet Gynecol 12:39–42
Morozov VN, Gavryushkin AV, Deev AA et al. (2002) Direct detection of isotopically labeled metabolites bound to a protein microarray using a charge-coupled device. J Biochem Biophys Methods 51:57–67
Ornstein DK, Gillespie JW, Paweletz CP et al. (2000) Proteomic analysis of laser capture microdissected human prostate cancer and in vitro prostate cell lines. Electrophoresis 21:2235–2242
Paweletz CP, Gillespie JW, Ornstein DK et al. (2000) Rapid protein display profiling of cancer progression directly from human tissue using a protein biochip. Drug Dev Res 49:34–42
Paweletz CP, Charboneau L, Bichsel VE et al. (2001) Reverse phase protein microarrays which capture disease progression show activation of pro-survival pathways at the cancer invasion front. Oncogene 20:1981–1989
Pawson T (1995) Protein modules and signalling networks. Nature 373:573–580
Petricoin EF, Liotta LA (2003) Mass spectrometry-based diagnostics: the upcoming revolution in disease detection. Clin Chem 49:533–534
Petricoin EF 3rd, Ornstein DK, Paweletz CP et al. (2002) Serum proteomic patterns for detection of prostate cancer. J Natl Cancer Inst 94:1576–1578
Petricoin EF, Ardekani AM, Hitt BA et al. (2002) Use of proteomic patterns in serum to identify ovarian cancer. Lancet 359:572–577
Petricoin EF, Zoon KC, Kohn EC et al. (2002) Clinical proteomics: translating benchside promise into bedside reality. Nat Rev Drug Discov 1:683–695
Rosty C, Christa L, Kuzdzal S et al. (2002) Identification of hepatocarcinoma-intestine-pancreas/pancreatitis-associated protein I as a biomarker for pancreatic ductal adenocarcinoma by protein biochip technology. Cancer Res 62:1868–1875
Schwartz SA, Weil RJ, Johnson MD et al. (2004) Protein profiling in brain tumors using mass spectrometry: feasibility of a new technique for the analysis of protein expression. Clin Cancer Res 10:981–987
Schweitzer B, Roberts S, Grimwade B et al. (2002) Multiplexed protein profiling on microarrays by rolling-circle amplification. Nat Biotechnol 20:359–365
Schwikowski B, Uetz P, Fields S et al. (2000) A network of protein–protein interactions in yeast. Nat Biotechnol 18:1257–1261
Shen Y, Tolic N, Zhao R et al. (2001) High-throughput proteomics using high-efficiency multiple-capillary liquid chromatography with on-line high-performance ESI FTICR mass spectrometry. Anal Chem 73:3011–3021

Shen Y, Tolic N, Masselon C et al. (2004) Ultrasensitive proteomics using high-efficiency on-line micro-SPE-nanoLC-nanoESI MS and MS/MS. Anal Chem 76:144–154

Simone NL, Remaley AT, Charboneau L et al. (2000) Sensitive immunoassay of tissue cell proteins procured by laser capture microdissection. Am J Pathol 156:445–452

Stillman BA, Tonkinson JL (2000) FAST slides: a novel surface for microarrays. Biotechniques 29:630–635

Stoeckli M, Chaurand P, Hallahan DE et al. (2001) Imaging mass spectrometry: a new technology for the analysis of protein expression in mammalian tissues. Nat Med 7:493–496

Templin MF, Stoll D, Schrenk M et al. (2002) Protein microarray technology. Trends Biotechnol 20:160–166

Tirumalai RS, Chan KC, Prieto DA et al. (2003) Characterization of the low molecular weight human serum proteome. Mol Cell Proteomics 2:1096–1103

Tonkinson JL, Stillman BA (2002) Nitrocellulose: a tried and true polymer finds utility as a post-genomic substrate. Front Biosci 7:c1–12

Tyers M, Mann M (2003) From genomics to proteomics. Nature 422:193–197

Warburg O, Posenor K, Negellein E et al. (1924). Biochem Z 152:309–345

Wiese R (2003) Analysis of several fluorescent detector molecules for protein microarray use. Luminescence 18:25–30

Wilson DS, Nock S (2003) Recent developments in protein microarray technology. Angew Chem Int Ed Engl 42:494–500

Wulfkuhle JD, McLean KC, Paweletz CP et al. (2001) New approaches to proteomic analysis of breast cancer. Proteomics 1:1205–1215

Wulfkuhle JD, Sgroi DC, Krutzsch H et al. (2002) Proteomics of human breast ductal carcinoma in situ. Cancer Res 62:6740–6749

Wulfkuhle JD, Aquino JA, Calvert VS et al. (2003) Signal pathway profiling of ovarian cancer from human tissue specimens using reverse-phase protein microarrays. Proteomics 3:2085–2090

Wulfkuhle JD, Liotta LA, Petricoin EF (2003) Proteomic applications for the early detection of cancer. Nat Rev Cancer 3:267–275

Xiao Z, Adam BL, Cazares LH et al. (2001) Quantitation of serum prostate-specific membrane antigen by a novel protein biochip immunoassay discriminates benign from malignant prostate disease. Cancer Res 61:6029–6033

Yanagisawa K, Shyr Y, Xu BJ et al. (2003) Proteomic patterns of tumour subsets in non-small-cell lung cancer. Lancet 362:433–439

Zhu H, Snyder M (2001) Protein arrays and microarrays. Curr Opin Chem Biol 5:40–45

Zhu H, Snyder M (2003) Protein chip technology. Curr Opin Chem Biol 7:55–63

11 Protocols for Tissue Handling and Molecular Analysis

This section of the book details the protocols for the preparation of tissue along with the analysis of DNA, RNA, and proteins from tissue samples. Most of these protocols are derived from our experience at the Pathogenetics Unit of the National Cancer Institute. Protocols contributed from other groups are specified.

11.1 Tissue Preparation for Prostate Prepared as Whole-Mount

11.1.1 70% Ethanol Fixation

Materials. 70% ethanol at 4 °C

Method

1. The fixation of tissue in 70% ethanol is similar for subsequent embedding in either low-melt polyester or paraffin.
2. Immediately after resection, the tissue specimen is placed on wet ice or into 70% ethanol at 4 °C.
   TIP: Placement into 70% ethanol may restrict subsequent processing, e.g., inking of surgical markings.
3. Transport to the pathology department for evaluation and further processing.
4. Grossly cut the specimen into thin sections to allow rapid penetration of fixative (maximum thickness of 3 mm).
   TIP: This is particularly important for molecular profiling studies because ethanol does not penetrate tissue as rapidly as formalin and it is critical to minimize the interval between surgical removal and cellular fixation.
5. Place in sufficiently larger container with a 10× volume of cold 70% ethanol.

Principles and Practice
Emmert-Buck et al. (Eds.):
Dissecting the Molecular Anatomy of Tissue

6. Fix overnight at 4 °C.
   TIP: Small biopsy samples can be fixed for shorter time intervals, typically 4–6 h.

11.1.2 Embedding

11.1.2.1 Low-Melt Polyester Embedding

Materials

1. 70, 90, 99, and 100% ethanol
2. Embedding molds, super metal-based, 66×54×15 mm (Surgipath Med. Ind. Inc.)
3. Embedding hardwood block, 1.5×1.5×1.0 in. (Shandon Lipshaw)
4. Microtome (Leica Microsystems Inc.)
5. Positively charged glass microscope slides (Brain Research Labs.)
6. Low-melt polyester (Gallard-Schlesinger Industries, Inc.)

Method

1. Large tissue specimens (several cm size)
   a) After tissue fixation in 70% ethanol, place tissue in fresh 70% ethanol at 4 °C for 2 h (×2)
   b) Place tissue in 90% ethanol at 4 °C for 1.5 h.
   c) Place tissue in 99% ethanol at 4 °C for 1.5 h.
   d) Place tissue in 100% ethanol at 25 °C for 2.5 h.
   e) Melt the low-melt polyester at 38 °C in an incubator and prepare two polyester : ethanol mixtures:

50:50,	polyester: 100% ethanol
90:10,	polyester: 100% ethanol

   TIP: Do not use a microwave for heating nor exceed 45 °C as these conditions will alter the properties of the polyester.
   f) Infiltrate tissue in the above 50:50 polyester:ethanol mixture at 45 °C for 2.5 h with agitation.
   g) Infiltrate tissue in the above 90:10 polyester:ethanol mixture at 45 °C overnight with agitation. Discard unused polyester after melting and do not reuse.
   h) Transfer tissue to an embedding mold in 90:10 polyester:ethanol mixture and chill until a pellicle has formed on the surface of the polyester adjoining the tissue.
   i) Trim the tissue block of excess polyester leaving a small margin around the tissue.

   j) Attach the tissue block to a hardwood block by melting the back of the tissue block with a warm spatula and firmly pressing the two together.
   k) Allow the tissue block to solidify at room temperature.
   l) Store tissue blocks in the refrigerator at 4 °C.
2. Small Tissue Specimens and Biopsies (mm to 2-cm size)
   a) After tissue fixation in 70% ethanol, place tissue in fresh 70% ethanol at 4 °C for 1 h
   b) Place tissue in 90% ethanol at 4 °C for 1 h.
   c) Place tissue in 99% ethanol at 4 °C for 1 h.
   d) Place tissue in 100% ethanol at 25 °C for 1 h.
   e) Infiltrate tissue in 50:50 polyester:ethanol mixture at 45 °C for 2 h with agitation.
   f) Infiltrate tissue in 90:10 polyester:ethanol mixture at 45 °C for 1 h with agitation.
   g) Transfer tissue to plastic embedding mold in 90% polyester:100% ethanol and chill on ice.
   h) Store tissue blocks in the refrigerator at 4 °C.
3. Histology Slide Cutting Procedure
   a) Use a standard microtome to cut 8-μm-thick sections.
   b) Place sections in deionized water bath at room temperature.
   c) Place the sections on positively charged glass slides and drain excess water.
   d) Dry slides for 5 min at 30 °C.
   e) Store slides at 4 °C until use.

11.1.2.2 Paraffin Embedding

Materials
1. Sakura FineTek V.I.P. tissue processor (Sakura FineTek Inc.)
2. 70, 80, 95, 100% ethanol
3. Paraffin wax (Paraplast; melting point 58 °C)
4. Xylenes, Reagent Grade (Sigma)
5. Embedding molds

Method
1. Fix tissue overnight in 70% ethanol at 4 °C.
2. Place the tissue in the tissue processor and process with the times and temperatures described in Table 11.1.
3. Embed the specimen in paraffin and block.
4. Cut sections into 5–8 μm thickness.
6. Both paraffin blocks and sections may be stored at or below room temperature.
   TIP: Tissue processors in standard histopathology laboratories generally include formalin fixation as the first two steps in the paraffin infiltration

Table 1. Conditions for processing tissues

Station	Solution	Concentration (%)	Time (min)	Temperature (°C)
1	Ethanol	70	30	40
2	Ethanol	80	30	40
3	Ethanol	95	45	40
4	Ethanol	95	45	40
5	Ethanol	100	45	40
6	Ethanol	100	45	40
7	Xylenes	100	45	40
8	Xylenes	100	45	40
9	Paraffin		30	58
10	Paraffin		30	58
11	Paraffin		30	58
12	Paraffin		30	58

procedure. Make certain these steps are avoided when processing tissue intended for molecular profiling.

11.2 Slide Preparation for Microdissection to Perform Subsequent DNA, RNA, and Protein Analysis

11.2.1 Slide Preparation for Laser Capture Microdissection (LCM)

Materials

1. 70, 95, and 100% ethanol
2. Xylenes, mixed, ACS reagent (Sigma)
3. Deionized water
4. Hematoxylin solution, Mayer's (Sigma)
5. Eosin Y solution (Sigma)
6. Complete, mini-protease inhibitor cocktail tablets (Roche Corp.)
   Important: For all protein analyses, dissolve one protease inhibitor cocktail tablet per 10 ml of each reagent except the xylenes.

Storage of Sections

1. Recut paraffin sections are stored at or below room temperature. Do not deparaffinize until immediately prior to microdissection.
2. Low-melt polyester sections are stored at 4 °C.
3. Frozen sections are stored at –80 °C or below.

Methods

TIP: Use the minimal amount of staining to visualize the tissue for microdissection. This will significantly improve macromolecule recovery. For example, hematoxylin and eosin can be used at 10% of their standard concentrations. Since the slides are microdissected without a coverslip, the tissue is not index-matched and substantial light scattering occurs, typically producing "dark" images. Thus, both image quality and molecular recovery can be improved by decreasing stain concentrations.

11.2.1.1 Paraffin-Embedded Sections and Frozen Sections

a) If a paraffin-embedded section is to be stained, start from step 1.
b) If the section was frozen-embedded, melt it gently (e.g., on the back of the hand) for approximately 30 s after removal from the freezer. This will create a "rougher" tissue surface and allow for better adhesion to the LCM cap. Start at step 5.
c) Place the sections in the following solutions:

1	Fresh xylenes (to deparaffinize the sections)	5 min
2	Fresh xylenes	5 min
3	100% Ethanol	15 s
4	95% Ethanol	15 s
5	70% Ethanol	15 s
6.	Deionized water	15 s
7	Mayer's hematoxylin	30 s
8	Deionized water, rinse (2×)	15 s
9	70% Ethanol	15 s
10	Eosin Y	5 s
11	95% Ethanol	15 s
12	95% Ethanol	15 s
13	100% Ethanol	15 s
14	100% Ethanol	15 s
15	Xylenes (to ensure dehydration of the section)	60 s
16	Air-dry for approximately 2 min or gently use air gun to completely remove xylenes	
17	The tissue is now ready for LCM	

11.2.1.2 Low-Melt Polyester-Embedded Sections

TIP: Proceed gently when staining sections embedded in polyester wax. Even though the sections are placed on charged slides, the tissue has a tendency to detach from the slide and should be monitored carefully throughout the staining procedure.

Place the sections in the following solutions:

1	100% Ethanol (to remove polyester wax)	5 min
2	100% Ethanol	5 min
3	95% Ethanol	15 s
4	70% Ethanol	15 s
5	Deionized water	15 s
6	Mayer's hematoxylin	30 s
7	Deionized water	15 s
8	70% Ethanol	30 s
9	Eosin Y	5 s
10	95% Ethanol	15 s
11	95% Ethanol	15 s
12	100% Ethanol	15 s
13	100% Ethanol	15 s
14	50:50, Xylenes:100% ethanol	10 s
15	The tissue is now ready for LCM	

TIP: The xylene-ethanol step at the end of the procedure is critical for subsequent LCM. The length of time may need to be adjusted depending on the tissue type and goals of the study. For example, if the tissue is left in this solution longer than 10–15 s, the tissue may detach from the slide during dissection. Conversely, if the xylene-ethanol step is too short, the tissue may be strongly bound to the slide and may not dissect well.

TIP: The tissue section should be completely dry before LCM. Use of an Accuduster or similar device may facilitate drying for efficient microdissection.

11.2.2 Slide Preparation for Manual Microdissection

Materials

1. 70, 95, and 100% ethanol
2. Xylenes, mixed, ACS reagent (Sigma)
3. Glycerol, ultra-pure (Gibco)
4. Deionized water
5. Hematoxylin solution, Mayer's (Sigma)
6. Eosin Y solution (Sigma)
7. Complete, mini-protease inhibitor cocktail tablets (Roche Corp.)
   Important: For all protein analysis, dissolve one protease inhibitor cocktail tablet per 10 ml of each reagent, except xylene.

Storage of Sections

1. Recut paraffin sections are stored at or below room temperature. Do not deparaffinize until immediately prior to microdissection.
2. Low-melt polyester sections are stored at 4 °C.
3. Frozen sections are stored at –80 °C or below.

Methods

TIP: Use the minimal amount of staining to visualize the tissue for microdissection. This will significantly improve macromolecule recovery. For example, hematoxylin and eosin can be used at 10% of their standard concentrations. Since the slides are microdissected without a coverslip, the tissue is not index-matched and substantial light scattering occurs, typically producing "dark" images. Thus, both image quality and molecular recovery can be improved by decreasing stain concentrations.

11.2.2.1 Paraffin-Embedded Sections and Frozen Sections

a) For paraffin-embedded sections, start at step 1.
b) For frozen-embedded sections, start at step 4.
c) Place the sections in the following solutions:

1	Fresh xylenes (to deparaffinize the sections)	5 min
2	Fresh xylenes	5 min
3	100% Ethanol	30 s
4	95% Ethanol	30 s
5	70% Ethanol	30 s
6	Deionized water	30 s
7	Mayer's Hematoxylin	30 s
8	Deionized water	Rinse 15 s (2×)
9	70% Ethanol	30 s
10	Eosin Y	15 s
11	Deionized water	30 s (2×)
12	3% Glycerol in deionized water	30 s
13	After removing the slide from the 3% glycerol step, shake the slide in the air to remove the layer of glycerol/water	

TIP: Soaking in 3% glycerol is particularly helpful in preparing the tissue for microdissection because it renders the tissue less brittle and dissected tissue fragments are easier to procure.

TIP: The next 5–10 min are the optimal time for microdissection. The tissue is dry, but retains a soft consistency. If the dissection takes more than a few minutes, the tissue will become increasingly brittle and the dissected fragments may be repelled as the needle is brought in proximity to the tissue. If the tissue becomes overly dry, resoak in the 3% glycerol/water solution for 1–2 min.

11.2.2.2 Low-Melt Polyester-Embedded Sections

TIP: Proceed gently when staining sections embedded in polyester wax. Even though the sections are placed on charged slides, the tissue has a tendency to detach from the slide and, therefore, should be monitored carefully throughout the staining procedure.

Place the sections in the following solutions:

1	100% Ethanol	5 min
2	100% Ethanol	5 min
3	95% Ethanol	30 s
4	70% Ethanol	30 s
5	Deionized water	30 s
6	Mayer's hematoxylin	30 s
7	Deionized water	30 s
8	70% Ethanol	30 s
9	Eosin Y	15 s
10	Deionized water	30 s (2×)
11	3% Glycerol in deionized water	30 s
12	The tissue is now ready for microdissection	

TIP: It is important to ensure that the thin coat of fluid covering the slide after removal from the glycerol/water is removed. Dissection with this fluid layer present results in diffusion of tissue fragments with potential to "contaminate" samples. In addition, dissection of the tissue under fluid produces large strips of tissue that are not easily homogenized in extraction buffers.

11.3 Techniques for Tissue Microdissection

11.3.1 Laser Capture Microdissection

Method

1. Once the tissue has been properly processed, sectioned, and stained, it is ready for microdissection.
2. The tissue section is placed on the stage of the laser microscope.
3. A cap with attached thermoplastic film is placed onto the tissue.
4. The tissue is visualized under the microscope and an initial road map image is taken.
5. The laser is fired repeatedly encompassing the cells to be dissected.
6. Pre-dissection image is taken.
7. The cap is lifted from the tissue removing the targeted cells from the tissue section.
8. A post-dissection image is taken.
9. The cap with dissected cells is imaged.
10. The cap is placed into an Eppendorf tube containing the appropriate buffer depending on the analysis being performed.

TIPS:

1. It is essential that there are no irregularities on the tissue surface or near the area to be microdissected. Wrinkles will elevate the LCM cap away from the tissue surface and decrease the membrane contact during laser activation.
2. Occasionally, there are subtle irregularities on the tissue surface (under the LCM cap) that cannot be visually appreciated; however, these can be noted by a decrease in the laser activation spot size. This can be partially or completely alleviated by adding an extra weight to the cap support arm, or temporarily increasing the laser power.
3. Use an adhesive pad after microdissection to remove cells that may have attached nonspecifically to the LCM cap. Place the cap on the adhesive pad three separate times and then view it microscopically to ensure that all of the nonspecific material has been removed.
4. A cap-alone control is recommended for each experiment to ensure that nonspecific transfer is not occurring during microdissection. This is best performed by placing an LCM cap on the tissue section being dissected and aiming and firing the laser at regions where there are no cells or structures present, e.g., lumens of large vessels, cystic structures, etc. (alternatively, one can place a portion of the LCM cap "off" the tissue and target this region). The cap should be processed through the buffer and analysis methodology being utilized in the study and serves as a negative control.

11.3.2 Immuno-Laser Capture Microdissection

Materials

1. 1× phosphate-buffered saline, pH 7.4
2. Acetone
3. 70, 95, and 100% ethanol
4. Xylenes, mixed, ACS grade (Sigma)
5. DAKO Quick Staining kit (DAKO Corp.), a three-step streptavidin-biotin technique with prediluted mono- or polyclonal (rabbit) primary antibodies optimized for very short staining times.
6. Diaminobenzidine, DAB
7. Hematoxylin solution, Mayer's (Sigma)
8. Superfrost Plus glass slides (Fisher Scientific)
9. Placental RNase inhibitor (Perkin Elmer, Branchburg)
10. DEPC-treated H_2O (Research Genetics)

    Important: Add placental RNase inhibitor to the primary antibody and the DAB solution in a concentration of 200–400 U/ml. All solutions are prepared with DEPC-treated water.

Method (under RNase-free conditions)

1. Cut 8-μm serial sections of snap-frozen tissue blocks on a standard cryostat with a new disposable cryostat blade.
2. Mount the tissue sections on Superfrost Plus glass slides and store immediately at −80 °C.
3. Thaw the frozen sections at room temperature for 30–60 s without drying.
4. Fix by immersing immediately in cold acetone for 5 min.
5. Rinse the slides briefly in 1× phosphate-buffered saline, pH 7.4.
6. Using the DAKO Quick Staining kit, immunostain the slides by incubating them at room temperature with the primary and secondary antibodies and the tertiary reagent for 90–120 s each, rinsing briefly with 1× PBS between each step.
7. Develop the color with diaminobenzidine (DAB) for 3–5 min and counterstain with hematoxylin for 15–30 s.
8. Dehydrate the sections sequentially in 70, 95, and 100% ethanol (15 s each) and xylenes (twice for 2 min each).
9. Air dry.
10. The immunostained sections are then ready for LCM.
    TIP: For primary antibodies other than the prediluted DAKO Quick Staining kit antibodies, the dilutions should be determined individually.

11.3.3 Manual Microdissection

Method

1. Dissection is performed on a standard inverted microscope using a 30 gauge needle on a syringe as the microdissecting tool.
   TIP: To stabilize the dissecting hand, the dissector should prop elbow on a solid surface adjacent to and at the same height as the stage of the microscope. It is helpful to rest the ulnar aspect of the dissecting hand on the stage of the microscope and move the needle into the microscopic field, a few millimeters above the tissue. In this way, the dissecting arm and hand can be rested on solid support surfaces.
2. While viewing the tissue through the microscope, the cell population of interest should be gently scraped with the needle. The dissected cells will become detached from the slide and form small dark clumps of tissue that can be collected on the needle by electrostatic attraction. Several small tissue fragments can be procured simultaneously. Collection of an initial fragment on the tip of the needle will assist in procuring subsequent tissue. The tip of the needle with the procured tissue fragments should be carefully placed into a small PCR tube containing the appropriate buffer. Gentle shaking of the tube will ensure the tissue detaches from the tip of the needle.

TIP: Pressing down on the shaft of the syringe to inject an air bubble into the extraction solution helps to detach the tissue from the needle and prevents any fragments from remaining lodged in the barrel of the needle.

3. Placement of frozen tissue sections directly on agarose-coated slides can be helpful in maintaining enzyme stability. In addition, the agarose gels can be prepared or soaked in custom buffers that will bathe the frozen section prior to and during the microdissection, e.g., pH, salt concentration, proteinase inhibitors, etc., can be varied specifically for a given enzyme. Some members of the NCI Group also prefer to use the agarose-coated slide microdissection approach for recovery of mRNA. Slides for microdissection are prepared by placing 200 μl of warm agarose on standard uncoated glass slides, covering with a glass coverslip, and allowing the gel to polymerize. The glass coverslip is removed from the slide and the frozen tissue section is immediately placed onto the agarose gel. The freshly cut section should be transferred directly from the cryostat to the agarose-coated slide.
4. Microdissection can be performed similar to the method described above. However, the dissector may find it easier to "tease" the tissue apart since the tissue remains bathed in the fluid from the gel and can be gently pulled apart. The tissue will also separate along tissue planes, e.g., stroma and epithelium will easily separate from each other. The dissected tissue can be gently picked up from the slide, or alternatively, the dissector can use the needle to physically cut the agarose and procure both the agarose and the tissue fragment together.

11.4 Processing of Microdissected Tissue for Molecular Analysis

11.4.1 DNA-Based Studies

11.4.1.1 Protocol from Pathogenetics Unit

1. More than 10,000 cells
   a) If the amount of microdissected material is substantial (>10,000 cells), then any of the standard procedures for isolating DNA are acceptable.
2. Less than 10,000 cells
   a) If the number of cells procured is minimal, e.g., dissection of pre-malignant lesions, then a simple proteinase K treatment prior to PCR is recommended.
   b) Procured cells are suspended in buffer containing 10 mM Tris-HCl, 1 mM ethylenediamine tetraacetic acid (EDTA), 1% Tween 20, and 0.1 mg/ml proteinase K, pH 8.0, and incubated overnight at 37 °C.

Longer incubation times, higher incubation temperatures, and higher concentrations of proteinase K have been reported to improve the quality of DNA recovered from fixed tissue sections. The sample is subsequently boiled for 8 min to inactivate the proteinase K and the DNA is ready for PCR analysis.

TIP: If a lesion or particular cell population containing very small numbers of cells is desired, it is recommended that the identical lesion from as many consecutive histologic recuts as possible be dissected to maximize the total number of cells procured.

TIP: There is no optimal number of cells that should be collected from a microdissection since results vary significantly depending on the tissue source. For frozen tissue approximately 20 cells/μl of extraction buffer is recommended as a good starting point; however, more cells per PCR reaction may be needed for DNA recovered from formalin-fixed, paraffin-embedded tissue.

3. Formalin-fixed, paraffin-embedded tissue
   a) PCR amplification of DNA recovered from standard formalin-fixed, paraffin-embedded tissue is possible and allows investigators to perform studies on archival patient material. Studies on processed tissue have the general advantages of abundant samples for study, high quality of histologic detail for studying dysplastic and pre-malignant lesions, and frequent availability of follow-up clinical information on patients. However, these studies are often technically challenging. The type of fixative used and the fixation time impact heavily on the quality of the DNA recovered after microdissection, thus PCR amplification signals may be widely disparate among samples in a study.

      TIP: If a sample does not initially give a PCR product, a tenfold dilution of the boiled sample may yield a strong PCR product due to dilution of tissue inhibitors of DNA polymerase.

      TIP: Utilization of PCR primer sets that produce products in the 150–200 bp range is optimal since the DNA is often cross-linked and/or fragmented, and larger products may not be reliably amplified.
4. Other fixatives
   a) Fixatives containing heavy metals or low pH should be avoided. Non-formalin-based fixatives generally result in recovery of better DNA and may become increasingly important for routine processing of tissue samples in the future as the field of molecular pathology evolves.
5. Timing
   a) Routine care in tissue processing is also helpful, including immediate processing of samples after surgery and slicing of tissue samples into thin sections to allow rapid penetration of fixative. Overfixation (>24 h) should be avoided.

11.4.1.2 DNA Extraction (Protocol from Wistuba)

NOTE: For proteinase K digestion to extract DNA from microdissected formalinfixed, paraffin embedded tissue (FEPET), there are several important issues which need to be addressed:

1. Estimate accurately the number of cells microdissected.
   TIP: Although fluorometrical assessment of DNA concentration is available for minute amounts of FFPET, DNA quantitation is a very difficult task if microdissection of small lesions have been performed.
2. Optimize the PCR protocol for the minimum number of cells that provides reproducible results.
3. The basic protocol for DNA extraction suggests a proteinase K concentration of 200 μg/ml, containing 20 mM Tris (pH 8.0), 1 mM EDTA (pH 8.0) and 0.5% Tween 20.
4. Use 5 μl of extraction buffer for each set of 200 microdissected cells. Incubate at 37 °C for 24–48 h and heat-inactivate proteinase K in boiling water bath for 10 min.
5. The use of DNA purification methods is optional, and they may cause a loss of DNA if small numbers of microdissected cells are analyzed. The most frequently used DNA purification procedures are: incubation with Chelex (25% w/v), ethanol precipitation with 7.5 ammonium-acetate (pH 7.4), buffer exchange using spin columns, organic extraction with Tris-saturated phenol-chloroform (1 : 1; pH 8.0), and a combination of methods (i.e., phenol-chloroform with ethanol precipitation and Chelex incubation).
6. Preservation of extracted DNA at 4 °C is preferred if it is to be used within days. For longer preservation, keep samples at –20 °C. DNA degradation is enhanced with frequent thawing.

11.4.2 RNA Isolation

11.4.2.1 Isolation of Total RNA from Microdissected Frozen Tissue

Recovery and analysis of RNA from frozen tissue sections can be achieved using slight modifications of standard methods. We obtain total RNA from microdissected samples using the Micro RNA Isolation kit (Stratagene, La Jolla, CA).

Materials
Stratagene Microisolation Kit.

Method

1. Before beginning, clean all pipettes with RNase away or a similar product.
2. Place the LCM cap in an Eppendorf tube containing 200 μl RNA denaturing buffer (GITC) and 1.6 μl 2-mercaptoethanol.

3. Invert several times over the course of 2 min to digest the tissue. Transfer the solution from the reagent tube to a sturdy RNase-free 1.5-ml tube.
4. Add 20 μl (0.1× volume) 2 M sodium acetate, pH 4.0.
5. Add 220 μl (1× volume) water-saturated phenol (take bottom layer of 1:1 mixture).
6. Add 60 μl (0.3× volume) chloroform-isoamyl alcohol.
7. Shake the tube vigorously for 15 s.
8. Place on wet ice for 15 min.
9. Centrifuge for 30 min at 4 °C to separate the aqueous and organic phases.
10. Transfer upper aqueous layer to a new tube.
    TIP: If any of the lower organic phase was accidentally transferred and may be contaminating the aqueous phase, this will interfere with the subsequent isopropanol precipitation. To remove any residual organic material from the aqueous layer, add one volume of 100% chloroform, mix well, and centrifuge for 10 min at 4 °C to separate the aqueous and organic phases. Transfer the upper layer to a new tube.
11. Add to aqueous layer, 1–2 μl glycogen (10 mg/ml) and 200–300 μl cold isopropanol (i.e., equal volume).
    TIP: Glycogen facilitates visualization of the pellet, which can be problematic when using small amounts of RNA.
12. Place samples at –80 °C for at least 30 min. They may also be left overnight.
    TIP: Before centrifuging, the tubes may need to be thawed slightly if they have solidified during the isopropanol precipitation.
13. Centrifuge for 30 min at 4 °C with cap hinges pointing outward so that the location of the pellet can be better predicted.
14. Remove the supernatant and wash with 300 μl cold 70% ethanol. Spin for 5 min at 4 °C.
15. Remove the supernatant.
16. Let the pellet air-dry on ice to remove any residual ethanol (Overdrying prevents the pellet from resuspending easily).
17. The pellet may be stored at –80 °C until use or proceed to DNase treatment (below).

Optional DNase Treatment

TIP: DNAse treatment is highly recommended for microdissected cells. Genomic DNA contamination is often problematic with these samples, possibly due to the small DNA fragments that are created during tissue processing and are difficult to purify from RNA.

1. To RNA pellet, add 15 μl DEPC water and 1 μl 20 U/μl RNase inhibitor (Perkin Elmer).
2. Gently mix by flicking until the pellet is dissolved.
3. Quickspin.

4. Add 2 μl 10×DNase buffer (GenHunter) and 2 μl 10 U/μl DNase I (GenHunter; 20 U total).
5. Incubate at 37 °C for 2 h.
6. Re-extract RNA by adding: 2 μl 2 M sodium acetate, pH 4.0, 22 μl water-saturated phenol, 6 μl chloroform-isoamyl alcohol.
7. Vortex vigorously for 15 s.
8. Place on wet ice for 15 min.
9. Centrifuge 10 min at 4 °C.
10. Transfer upper layer to a new tube.
    TIP: If any of the lower organic phase was accidentally transferred and may be contaminating the aqueous phase, this will interfere with the subsequent isopropanol precipitation. To remove any residual organic material from the aqueous layer, add one volume of 100% chloroform, mix well, and centrifuge for 10 min at 4 °C to separate the aqueous and organic phases. Transfer the upper layer to a new tube.
11. Continue with RNA extraction from step 10 in Total RNA Isolation, adjusting the volume of isopropanol accordingly.

11.4.2.2 Protocol for Isolation of RNA from Frozen LCM Tissue (Protocol from Watson)

1. Prepare lysis buffer by adding 7.2 μl of 14.4 M 2-mercaptoethanol to 1 ml of denaturation solution (4 M guanidine isothiocyanate/20 mM sodium citrate/0.5% sarkosyl). Mix well.
2. Tissue microdissection is performed using the Pixcell II LCM instrument (Arcturus) and CapSure LCM transfer films. When all target cells within the transfer film area have been captured, remove the cap and place it in a 0.5-ml microcentrifuge tube containing 100 μl of lysis buffer. Invert the tube so that the lysis buffer covers the cap and incubate at room temperature for at least 10 min to allow tissue digestion off the cap.
3. While tissue captured on the first transfer film is being lysed, place a new film over an additional area of the section or on a new slide. Repeat laser capture of the desired cell populations.
4. Briefly spin the first cap and tube to collect the lysis buffer and lysed tissue. Remove the first cap and replace it with a new cap of dissected tissue. In this manner, multiple dissections of the same target cell population may be pooled into a single volume of lysis buffer.
5. After dissecting all areas of interest, scrape the remaining tissue from the slide into a tube containing 100 μl of lysis buffer to serve as a positive control for RNA isolation ('Tissue Control').
6. Transfer the tissue lysate from the 0.5-ml tube to a 1.5-ml tube. Add 2 μl of nuclease-free water and 10 μl of 2 M sodium acetate (pH 4.0). Vortex briefly. Add 100 μl water-saturated phenol and vortex briefly.

Add 20 μl chloroform:isoamyl alcohol (24:1) and vortex vigorously for 1 min.

7. Place on wet ice for 15 min and then centrifuge at 12,000×g for 5 min at room temperature to separate the aqueous and organic phases.
8. Transfer the upper aqueous layer (~100 μl) to a new 1.5-ml tube. Add 2 μl glycogen (10 mg/ml) as a carrier to the aqueous layer and mix well. Add 100 μl cold isopropanol and mix well. Place in a –20 °C freezer for at least 30 min. Samples may be stored indefinitely in isopropanol.
9. Centrifuge at 16,000–20,000×g for 30 min at 4 °C to precipitate RNA. Carefully wash RNA pellet with 400 μl of chilled 70% ethanol. Centrifuge at 16,000–20,000×g for 5 min at 4 °C and then remove as much supernatant as possible without disturbing the pellet. Allow the pellet to air dry for 5–10 min. Be certain that no ethanol is remaining in the tube (although a small amount of water may be present). Do not allow the pellet to over-dry as the precipitated RNA pellet may become difficult to resuspend in water. Resuspend the RNA pellet in 10 μl of nuclease-free water. Do not resuspend the pellet in DEPC-treated water, Tris-EDTA, or RNA stabilization buffers, as these may inhibit downstream enzymatic reactions. Store the RNA at –80 °C.
10. Note that treatment with DNase is not necessary for microarray analysis and, in fact, may result in significant degradation or total loss of RNA. Treatment with DNase is also not necessary for quantitative RT-PCR, assuming that primers can be designed to span across exon boundaries. In our experience, negative control reactions consisting of cDNA synthesis in the absence of reverse transcriptase enzyme (-RT control) seldom yield amplification signals before a cycle threshold (C_T) of 38 cycles, indicating that amplification of contaminating genomic DNA is usually not problematic.
11. Generation of high quality gene expression microarray data is dependent upon high quality input RNA. Therefore, whenever possible, RNA isolated from microdissected tissue should be directly assessed by UV spectroscopy and/or electrophoresis. The Nanodrop ND-100 fiberoptic spectrophotometer (Nanodrop Technologies, Rockland, DE) provides a convenient means to quantify 1 μl of isolated RNA with nanogram sensitivity. We also routinely use the Agilent 2100 Bioanalyzer (Agilent, Palo Alto, CA) with RNA Nano or RNA Pico Chip microcapillary gel electrophoresis assays. This system allows qualitative assessment and some degree of quantification of nanogram or picogram quantities of RNA, respectively. It also provides a convenient means to digitally archive electrophoretic images of RNA samples. However, as this approach requires specialized and relatively expensive instrumentation, other methods of RNA assessment may be more acceptable. For example, small amounts of RNA may be quantified by dye-binding fluorescence assays (e.g., RiboGreen assay, Molecular Probes, Eugene, OR).

Small amounts of RNA may also be qualitatively analyzed by mini-agarose gel electrophoresis followed by staining with sensitive fluorescent dyes (e.g., SybrGold dye, Molecular Probes, Eugene, OR). Although less convenient, this approach has been successfully utilized to qualitatively assess nanogram quantities of total RNA from microdissected tissue (Kabbarah et al. 2003 from references Chapter 9). If too few cells have been captured to analyze the microdissected RNA directly, RNA isolated from the remaining dissected tissue section ('Tissue Control' – see step 5 above) should be assessed.

11.4.2.3 Protocol for Isolation of RNA from Formalin-Fixed, Paraffin-Embedded LCM Tissue (Protocol from Watson)

NOTE: Several methods have been published for the isolation of RNA from FFPE tissue (Krafft et al. 1997; Coombs et al. 1999; Macabeo-Ong et al. 2002; Korbler et al. 2003 from references Chapter 9). These methods involve optional pretreatment with proteinase K for varying lengths of time and incubation temperature and/or organic extraction followed by precipitation. The results of these studies provide conflicting data with regard to the best approach for optimizing RNA yield and quality. Therefore, we routinely use a commercial reagent system from Ambion (Austin, TX), slightly modified for processing of LCM FFPE tissue.

1. Add 20 µl of Ambion proteinase K digestion buffer and 1 µl of proteinase K to a 0.5-ml tube. Attach the LCM cap with tissue to the tube. Invert the tube to allow the digestion buffer to cover the cap and tissue. Incubate the inverted tube at 45 °C for 15–30 min. Then, briefly spin the tube to collect the tissue lysate and remove LCM cap. Multiple caps from one dissection may be pooled into the same volume of lysis buffer as discussed in Sect. 11.4.2.2 above.
2. Add 120 µl of Ambion RNA extraction buffer to the tissue lysate, vortex briefly, and transfer the resulting 140-µl volume to a 1.5-ml tube. Incubate at room temperature for 5 min and then add 140 µl of acidic phenol/chloroform to the sample.
3. Vortex, incubate at room temperature for 5 min, and spin at room temperature for 15 min at 12,000×*g*. Transfer the aqueous (upper) layer to a new 1.5-ml tube.
4. Add 1 µl of linear acrylamide (5 mg/ml) to the aqueous sample and vortex. Then add an equal volume (~140 µl) of isopropanol to the sample and vortex.
5. Incubate at –20 °C for 30 min to overnight and then spin at >16,000×*g* for 30 min at 4 °C.
6. Carefully decant the supernatant and wash the pellet (which may be invisible) with 500 µl of nuclease-free 70% ethanol. Spin again at >16,000×*g* for 5 min at 4 °C.

7. Carefully decant the supernatant and allow the pellet to air dry for no more than 5–10 min. Resuspend the pellet in 10 μl of nuclease-free water.
8. 1 μl of RNA may be quantified and qualitatively assessed as discussed above in Section 11.4.2.2.

11.4.3 Protein-Based Studies

11.4.3.1 Sample Preparation for 2-D PAGE

1. To lyse a 5–8 μm tissue section obtained from a paraffin-embedded block
   a) Place the tissue section in a 1.5-ml Eppendorf tube.
   b) Add xylenes to cover the tissue.
   c) Vortex vigorously for ~15 s.
   d) Incubate at RT for 5 min.
   e) Vortex again.
   f) Spin down for 3 min to pellet the tissue.
   g) Remove xylenes.
   h) Add 1 ml xylenes.
   i) Vortex 5–10 s.
   j) Spin down.
   k) Remove xylenes.
   l) Speed vacuum the sample for a few minutes to evaporate the remaining xylene.
   m) Add 400 μl isoelectric focusing buffer (IEF).
   n) Vortex vigorously for 1 min.
   o) Incubate for 5 min at RT.
   p) Vortex vigorously for 1 min.
   q) Spin sample down at 14,000×*g* for 5–10 min at RT.
2. To lyse a 5–8 μm tissue section obtained from a polyester-embedded block
   a) Follow steps a–o above for a paraffin-embedded block section, EXCEPT use 100% ethanol instead of xylenes.
3. To lyse a 5–8 μm frozen tissue section OR a microdissected tissue sample
   a) Place the tissue section or microdissected tissue sample in a 1.5-ml Eppendorf tube.
   b) Add 400 μl IEF buffer.
   c) Vortex vigorously for 1 min.
   d) Incubate for 5 min at RT.
   e) Vortex vigorously for 1 min.
   f) Spin sample down at 14,000×*g* for 5–10 min at RT.
4. To lyse a paraffin-embedded section on a slide

a) Deparaffinize the tissue by immersing the slide into xylenes, twice for 5 min each.
b) Allow the tissue section to dry.
c) Add 200 μl of the IEF buffer to the tissue and pipette up and down several times.
d) Remove the buffer into a microfuge tube.
e) Scrape the tissue with a razor blade and transfer into the same microfuge tube.
f) Add 200 μl to the lysates for a total of 400 μl.
g) Vortex vigorously for 1 min.
h) Incubate for 5 min at RT.
i) Vortex vigorously for 1 min.
j) Spin sample down at 14,000×*g* for 5–10 min at RT.

5. To lyse a polyester-embedded section on a slide
   a) Remove the polyester by immersing the slide into ethanol, twice for 5 min each.
   b) Repeat steps b–j, directly above, as for a paraffin-embedded section on a slide.
6. To lyse a frozen section on a slide
   a) Thaw the slide to room temperature.
   b) Immerse the tissue section into xylenes for 1 min.
   c) Repeat steps b–j, directly above, as for a paraffin-embedded section on a slide.

11.4.3.2 Immunoblot

1. Staining and laser capture microdissection
   a) Place a pellet of the protease inhibitor cocktail into each of the staining dishes.
   b) Stain the 8–12 μm frozen section via the traditional LCM H & E staining protocol.
      TIP: Hematoxylin staining decreases subsequent protein recovery and should be avoided if possible. If hematoxylin is absolutely required for visualization during microdissection then approximately twice as many cells should be utilized.
   c) Microdissect 1000–2000 cells from the stained frozen section. Limit time spent microdissecting to 30 min or less per slide to minimize protein degradation.
      TIP: For lower abundance proteins, more cells may be required for protein detection.

11.5 DNA, RNA, and Protein Analysis

11.5.1 DNA Analysis

11.5.1.1 Clonality X-Chromosome Inactivation

11.5.1.1.1 Protocol from Pathogenetics Unit

Materials

1. DNA sample (see Sect. 11.4.1).
2. Proteinase K (Sigma).
3. Proteinase K buffer (0.05 M Tris-HCL, 0.001 M EDTA, 1% Tween 20, 0.1 mg/ml proteinase K, pH 8.0) Phenol:chloroform:isoamyl alcohol (1:1:1, v:v:v; Gibco).
4. 3 M Na acetate, pH 5.2 (Life Technologies).
5. Isopropanol.
6. Glycogen, 10 mg/ml (GenHunter).
7. 70% ethanol.
8. Buffer 1 (50 mM Tris-HCl, pH 8.0, 10 mM $MgCl_2$, 50 mM NaCl)
9. Hha I restriction enzyme (Life Technologies).
10. REact 2 buffer (Life Technologies).
11. 10× PCR buffer (100 mM Tris-HCl pH 8.3, 500 mM KCL, 15 mM $MgCl_2$, 0.01% w/v gelatine, autoclaved; Perkin Elmer).
12. dNTP, each 10 mM (Perkin Elmer).
13. 7-deaza dGTP, 10 mM (Boehringer Mannheim).
14. DMSO, minimum 99.5%, GC (Sigma).
15. Forward and reverse primers, each 20 μM (Human Androgen Receptor Gene (HUMARA) CAG trinucleotide repeat microsatellite primers are described in Allen et. al. (1992: Am J Hum Genet 51(6):1229–1239)
16. AmpliTaq Gold, 5 U/ml (Perkin Elmer).
17. 32^{P}-dCTP, 20 μCi/μl (NEN).
18. Gel-Mix 6 (Life Technologies).
19. 10× TBE (1 M Tris-HCl, 0.9 M Boric acid, and 0.01 M EDTA; Life Technologies).
20. Loading dye (95% formamide, 20 mM EDTA, 0.05% bromophenol blue, 0.05% xylene cyanol).

Methods

TIP: Investigators must be especially careful when using this methodology to analyze archival tissue specimens. Formalin-fixation in particular results in DNA that is difficult to amplify and often produces inconsistent PCR results, including artifactual allelic loss and poor amplification of large products. If this technique is utilized for analysis of archival samples, we highly recommend that replicate experiments (multiple independent dissections, triplicate PCR reactions, etc.) be used to verify results.

1. LCM and proteinase K treatment
   a) Obtain microdissected cells using the LCM procedure.
      TIP: The number of cells needed to successfully perform the assay varies depending on the quality and processing conditions of the tissue samples. 1000 cells is recommended as a good starting point.
   b) Suspend approximately 1000 microdissected cells in 20 μl proteinase K buffer.
   c) Incubate overnight at 37 °C.
2. DNA purification
   a) Lyse the cells overnight at 42 °C, in 50 μl phenol:chloroform:isoamyl alcohol.
   b) Boil 10 min at 95 °C.
   c) Vortex 1 min.
   d) Centrifuge 20 min at 14,000 rpm at RT.
   e) Pipette upper phase into new tube and discard lower phase.
   f) Add 5 μl 3 M NaAc.
   g) Add 250 μl isopropanol.
   h) Add 2 μl glycogen.
   i) Vortex briefly.
   j) Place on dry ice for 30 min.
   k) Centrifuge 14,000 rpm for 30 min.
   l) Add 300 μl 70% ethanol.
   m) Vortex briefly.
   n) Centrifuge 2 min at 14,000 rpm at RT.
   o) Discard supernatant.
   p) Dry pellet at RT for 5 min, or 37 °C for 2 min.
   q) Resuspend the pellet in 20 μl buffer 1.
   r) Incubate at 37 °C for 3 min to completely dissolve the pellet.
3. DNA digestion
   a) Pipette 8 μl of the resuspended DNA into a reaction tube.
   b) Add 1 μl REact buffer 2.
   c) Add 1 μl Hha I.
   d) Incubate overnight at 37 °C.
   e) Incubate at 90 °C for 10 min.
   f) Use 2 μl of this digested DNA for PCR (see below).
   g) Use 2 μl of the non-digested DNA from step 2r above as a negative control for PCR.
      TIP: Parallel analysis of control DNA that is known to be from a monoclonal population is recommended to verify the efficiency of the restriction digest. DNA recovered from a tissue specimen that was processed similar to the tissue sample under study is ideal.
4. PCR

Set up the following PCR reaction:

2.0 μl	DNA sample
2.0 μl	10× PCR buffer
0.4 μl	dATP
0.4 μl	dCTP
0.4 μl	dTTP
0.4 μl	7-deaza dGTP
0.4 μl	Forward primer
0.4 μl	Reverse primer
0.2 μl	AmpliTaq Gold
0.4 μl	32^{P} dCTP
1.0 μl	DMSO
12 μl	DEPC-H_2O
Total volume = 20 μl	

Cycles are as follows:

Cycles	Temperature (°C)	Time
1	94	20 s
40	94	45 s
	60	45 s
	72	2 min
1	72	20 min

5. Gel electrophoresis
   a) Prepare gel consisting of 6% acrylamide as in LOH protocol.
   b) Add 4 μl dye to 20 μl PCR product.
   c) Denature the samples for 5 min at 94 °C.
   d) Load 3 μl onto gel.
   e) Electrophorese at 1800 V for 1–2 h.
   TIP: In clonality analysis, the PCR products are ~200 bp. Therefore, do not overload the gel with sample and run the gel for a longer period of time to sharpen the bands and increase separation.
   f) Transfer the gel to 3 mm Whatman paper and dry.
   g) Perform an autoradiograph with Kodak BIOMAX film as described in LOH protocol.
   TIP: Try a short exposure (1 h) time first and only re-expose depending on the results.
6. Interpretation of results
   A sample is considered to be composed of a monoclonal cell population when two alleles are recognized in the undigested DNA sample and complete absence of one allele is seen in the digested DNA sample.
   TIP: DNA that is recovered from microdissected samples and "semi-purified" using a one-step proteinase K buffer strategy will sometimes produce "non-specific" PCR products in addition to the true alleles. Nor-

mal-cell DNA from archival specimens from control subjects serves as a good comparator and can assist in interpretation of allele patterns.

11.5.1.1.2 HUMARA Assay (Protocol from Wistuba)

NOTE: The most important technical issues for X-chromosome inactivation analysis to assess clonality based on *HUMARA* assay are (Jovanovic et al. 2003 from references Chapter 8):

1. Proteinase K denaturation after DNA extraction has to be done at 75 °C, so proteinase K is denatured while DNA is maintained in double-stranded form for subsequent restriction digestions.
2. DNA purification has been demonstrated to be important to greatly improve restriction digestion.
3. PCR amplifiability of DNA needs to be addressed before restriction endonuclease digestion.
4. Between 200 and 450 ng of DNA are incubated with 40 U *Hha*I, 40 U *Hpa*II, and 40 U *Msp*I. The latter is an isoschizomer of *Hpa*II, which cleaves the CCGG restriction site within the *HUMARA* irrespective of the methylation status of the internal C residue, thus serving as a positive digestion control.
5. In all restriction endonuclease digestion experiments, negative and positive control reactions without enzymes and positive control reactions using both high-quality male reference DNA and bacterial DNA are strongly suggested.
6. Restriction enzyme incubation is recommended overnight at 37 °C, and the reaction is terminated by heat-inactivating the enzyme at 70 °C for 20 min.
7. For *HUMARA* PCR amplification, the following primer sequences are recommended: TCCAGAATCTGTTCCAGAGCGTGC (sense) and GCTGTGAAGGTTGCTGTTCCTCAT (antisense).
8. For FFPETs, touch-down PCR protocol has been suggested. These cycling conditions include 7 cycles with annealing at 67 °C, followed by 5 cycles at 65 °C and 28 cycles at 63 °C.
9. The methods for amplification visualization include gel electrophoresis using radioactive isotopes or gel silver staining.

11.5.1.2 Loss of Heterozygosity (LOH)

Reagents

1. DNA sample (see Chap. 11.4.1, Processing of microdissected tissue-DNA-based analysis)
2. Proteinase K (Sigma)
3. Proteinase K buffer (0.05 M Tris-HCl, 0.001 M EDTA, 1% Tween 20, 0.1 mg/ml proteinase K, pH 8.0)
4. AmpliTaq Gold Buffer (Perkin Elmer)

5. dNTP mixture (Perkin Elmer)
6. Primers
7. DEPC-treated H_2O
8. AmpliTaq Gold Polymerase (Perkin Elmer)
9. a-32^P dCTP, 6000 Ci/mmol (NEN Dupont)
10. Formamide, 99% (Fluka)
11. Bromophenol blue-xylene cyanole (Sigma), reconstituted as directed in Gel Mix-6 sequencing gel solution (Life Technologies)
12. Ammonium persulfate (Biorad)
13. 10×TBE buffer (0.89 M Tris base, 0.89 M Boric acid 0.02 M Disodium EDTA; Advanced Biotechnologies)
14. Acrylease (Stratagene)
15. Glass cleaner (e.g., Windex, Glass Plus)
16. 95% ethanol

Equipment
1. Thermal cycler (MJ Research)
2. Sequencing gel electrophoresis apparatus (Gibco BRL)
3. High voltage power supply
4. Gel dryer (Life Technologies)
5. Glass plates, 31.0×38.5 cm (Life Technologies)
6. 0.4-mm spacers (Life Technologies)
7. Casting boot (Life Technologies)
8. Shark tooth comb (Life Technologies)
9. Small clamps
10. Whatman blotting paper, 3 mm thickness
11. Kodak Biomax MR or AR film
12. Film cassette (Amersham Life Science)
13. Film processor

Time Requirements
1. Gel preparation: 1.5–2 h. Polymerization requires 1 h, but may stand overnight.
2. LOH reactions: 2.5 h (approximately 1 h for setup, 1.5 h for PCR)
3. High-resolution denaturing polyacrylamide gel electrophoresis: 1–3 h; 20 min for setup.
4. Electrophoresis time varies according to product size.
5. Gel drying: 1 h
6. Autoradiography: 1 h–2 days

Methods

TIP: Investigators must be especially careful when using this methodology to analyze archival tissue specimens. Formalin fixation in particular results in DNA that is difficult to amplify and often produces inconsistent PCR results, including artifactual allelic loss and poor amplification of large products. Therefore, when this technique is used to analyze archival samples, it is highly recommended that replicate experiments (multiple independent dissections, triplicate PCR reactions, etc.) be used to verify results.

1. LCM and proteinase K treatment
   a) Obtain microdissected cells using the LCM procedure.
   TIP: The number of cells needed to successfully perform the assay varies depending on the quality and processing conditions of the tissue samples. 1000 cells are recommended as a good starting point.
   b) Suspend approximately 1000 microdissected cells in 20 μl proteinase K buffer.
   c) Incubate overnight at 37 °C.
2. Prepare the glass plates
   TIP: Use Accuwipes for cleaning purposes, as they will not leave lint behind and are nonabrasive.
   a) Clean glass plates twice with glass cleaner.
   b) Repeat using 95% ethanol.
   c) Spray small plate with Acrylease.
   d) Spread Acrylease evenly using a circular motion.
   e) Buff dry.
   f) Quickly assemble the plates without touching the clean surface.
   g) Place 0.4-mm spacers on the edges of the larger plate.
   h) Place the smaller glass plate on top of the larger plate and spacers.
   i) Secure the plates with a casting boot (tape or clamps may be substituted for the casting boot).
3. Polymerize the gel
   TIP: Acrylamide is a neurotoxin. Be sure to wear gloves and a lab coat when working with this substance.
   a) Add 480 μl of 10% ammonium persulfate to 75 ml of Gel-mix-6.
   b) Mix by inversion.
   c) Hold the nozzle of the bottle at the corner of the gel cast.
   d) Hold the gel cast at a 45° angle to the bench and pour the gel between the plates (if bubbles get trapped between the plates, remove them by tapping the outside of the plates or by tipping the plates upright).
   e) Insert the straight side of the comb approximately 1 cm into the gel (if bubbles are introduced at this point, remove the comb and use the teeth of the comb to sweep out small bubbles).
   f) Clamp the top of the plates together.
   g) Allow the gel to polymerize for at least 1 h.
   TIP: The gel can be left to polymerize overnight. However, if bubbles appear, the gel has begun to separate from the plates. To minimize separation, wrap the gel in plastic film and store at 4 °C until use.
4. PCR reaction
   TIP: Investigators must be especially careful when using this methodology to analyze archival tissue specimens. Formalin fixation in particular results in DNA that is difficult to amplify and often produces inconsistent PCR re-

sults, including artifactual allelic loss and poor amplification of large products. If this technique is to be utilized for analysis of archival samples, we highly recommend that replicate experiments (multiple independent dissections, triplicate PCR reactions, etc.) be used to verify results.

a) Remove reagents from the freezer before beginning the procedure.
b) Thaw thoroughly before use.
c) Prepare all reactions on ice.
d) Prepare the reduced cytosine mixture prior to beginning the LOH reaction setup.
e) Vortex all reagents, with the exception of Taq Gold Polymerase before beginning the PCR reaction setup.
f) Prepare 320 µl reduced nucleotide mixture:

Reduced nucleotide mix

10 µl	dATP,	10 mM
10 µl	dGTP,	10 mM
10 µl	dTTP,	10 mM
2.0 µl	dCTP,	10 mM
288 µl	DEPC-treated H_2O	

g) Aliquot 1 µl of each DNA sample into a separate PCR tube and set aside.
   TIP: DNA that is recovered from microdissected samples and "semi-purified" using a one-step proteinase K buffer will sometimes produce "nonspecific" PCR products in addition to the true alleles. Moreover, larger alleles will sometimes amplify much less well than smaller alleles. Thus, normal-cell DNA recovered from the same tissue section as the tumor DNA serves as the best control for determining the presence or absence of allelic loss.
h) Prepare sufficient volume of the reaction mixture in a separate tube for all reaction tubes:

Reaction mixture/reaction tube

1.0 µl	Taq Buffer
0.8 µl	Reduced nucleotide mixture
0.2 µl	Forward primer, 20 µM
0.2 µl	Reverse primer, 20 µM
6.6 µl	DEPC-treated H_2O
0.1 µl	a-32^P dCTP
0.1 µl	AmpliTaq Gold Polymerase
Total volume=9 µl	

i) Thoroughly mix the reaction mixture by pipetting and dispense 9 µl of the reaction mixture into each tube containing DNA sample.

TIP: Be sure to mix the LOH reaction mixture with the DNA sample by pipetting. This is especially critical for DNA from microdissected samples that has been processed through a one-step proteinase K-based "purification."

j) Cap the reaction tubes and place them in a thermal cycler.
k) Cycle the reactions according to Tm of the specific primer set.
l) After PCR, remove the samples from the thermal cycler and dispense 2 µl of formamide/dye solution (95% formamide, 20 mM EDTA, 0.05% bromophenol blue, 0.05% xylene cyanole) into each reaction tube.
m) Store reactions at 4 °C until the gel is ready for loading.
   TIP: Investigators may want to consider the "touchdown" procedure for PCR by Don et al. (1991: Nucleic Acids Res 19:4008). Advantages include much cleaner bands, since by starting with a high annealing temperature of 66 °C and lowering 1 °C every cycle, the first PCR products are the most specific ones. Exactly the same protocol can be used for all primers.

5. Finalize gel preparation
   a) Remove the gel from the casting boot.
   b) Push the spacers into the gel until they are flush with the smaller glass plate to prevent the buffer from leaking during electrophoresis (spacers tend to get pushed out of the gel during polymerization).
   c) Place the gel in the sequencing apparatus and close the buffer release valve.
   d) Pour 500 ml of 0.5×TBE buffer into the upper chamber and 500 ml of 1×TBE buffer into the lower chamber.
   e) Remove the comb and clear bubbles from the loading area with a pipette.
   f) Insert the teeth of the comb approximately 1 mm into the gel.
   g) Pre-heat the gel at 1700 V for 15–20 min.
6. Gel loading
   a) Remove the samples from the freezer.
   b) Denature the samples in a thermal cycler at 95 °C for 5 min.
   c) Remove samples from the thermal cycler and immediately place on ice, with an ice pack on top of the samples, for 1 min.
   d) Turn off the power supply.
   e) Adjust comb if it has been pushed out of the gel during pre-heating.
   f) Load 4 µl of each sample per well.
      TIP: It is best to skip lanes to avoid contamination caused by leaking between the wells.
   g) Run the gel at 1700 V for 1–2 h (running time based on PCR product size).
7. Separate the gel
   a) Turn off the power supply.
   b) Drain buffer chambers (buffer must be disposed of in a liquid radioactive waste carboy).

c) Remove the gel from the sequencing apparatus.
d) Separate the plates by removing the spacers and inserting the tips of two thin spatulas in their place.
e) Gently lift the spatulas until the top plate separates from the lower plate and gel.
f) Place Whatman paper on the gel.
g) Slowly peel the Whatman paper and gel off the glass plate.
h) Cover the gel with plastic wrap and dry on a gel dryer for 1 h.

8. Autoradiography
   a) Remove the plastic wrap from the gel.
   b) Place the gel in an autoradiography cassette.
   c) Expose film 1 h–2 days using Kodak BioMax MR or AR film.
   TIP: Use MR film for maximum resolution of bands. An intensifying screen is useful when analyzing PCR products from small numbers of microdissected cells.

11.5.1.3 DNA PCR Amplification (Protocol from Wistuba)

NOTE: There is abundant scientific literature regarding PCR amplification of DNA from formalin-fixed, paraffin-embedded tissue, therefore few basic practical key aspects of this procedure are summarized:

1. Quality of the DNA extracted from FFPETs for PCR amplification depends on number of cells microdissected, and characteristics and duration of fixation (buffered formalin and no more than 24–48 h is desirable). If archival material is examined, a test for PCR amplification of the DNA should be performed using as a target a standard DNA sequence (i.e., β-goblin gene sequence; Sato et al. 2001 from references Chapter 8).
2. There are different types of PCR approaches available. Due to the small amount and the poor quality of DNA available, a two-round PCR approach is suggested. To maximize the use of the minute amounts of DNA obtained from microdissection, especially from preneoplastic lesions, multiplex PCR can be performed in the first round. In a multiplex PCR, several sequences are amplified (6–8) during the same reaction PCR, usually followed by uniplex PCR amplifying individual sequences (Peng et al. 1994 from references Chapter 8). In addition, nested or semi-nested designed primers can be utilized to optimize amplifications in a two-round PCR setting (Sugio et al. 1994; Kishimoto et al. 1995 from references Chapter 8).
3. As it has been established that formalin fixation and paraffin embedding produce fragmentation of the DNA in small fragments, primers have to be selected and designed for a PCR amplicon smaller than 250 bp (Legrand et al. 2002 from references Chapter 8).

11.5.1.4
Analysis of Microsatellite Markers (Protocol from Wistuba)

The basic PCR reaction used by us for amplification and electrophoresis analysis of polymorphic microsatellite DNA sequences in our LOH and MSI studies is summarized:

1. Nested PCR or two-round PCR (using the same set of primers in two consecutive amplifications) methods can be used. Multiplex PCR is performed during the first amplification, followed by a dilution step and consecutive uniplex PCR for an individual set of primers. In a multiplex PCR, several microsatellite sequences are amplified (6–8) during the same reaction.
2. If a multiplex PCR reaction is utilized, "hot-start" and "touch-down" PCR methodologies are recommended. We have obtained the best multiplex amplification results by using Taq polymerase systems that include a "hot-start" PCR approach (Wistuba et al. 1998; Wistuba et al. 2001 from references Chapter 8).
3. Volumes of 50 μl may be used for each multiplex reaction, containing 20 mM Tris (pH 8.3), 50 mM KCl, 2.5 mM $MgCl_2$, 400 μM of each deoxynucleotide triphosphate (dATP, dCTP, dGTP, dTTP), 0.5 μM of each forward and reverse primer, and 3.5 units of Taq polymerase.
4. The minimum number of microdissected formalin-fixed and paraffin-embedded cells needed for successful amplification of several markers in multiplex PCR is 100–200.
5. "Touch-down" PCR, which consists of decreasing the annealing temperature in the first cycles (i.e., cycles 3–10), allows the combination of several primer sets with different ranges of annealing temperatures in a single (multiplex) reaction (Don et al. 1991; Wistuba et al. 1998 from references Chapter 8). For the multiplex PCR reaction, a "touch-down" PCR is strongly recommended. After the initial denaturation at 94 °C for 10–12 min (depending on Taq polymerase system used), several (i.e., 11) cycles each consisting of denaturation at 95 °C for 20 s, annealing at 65–56 °C for 55 s and extension at 72 °C for 20 s can be performed, followed by an additional 24 cycles which include denaturation at 90 °C for 20 s, annealing at 55 °C for 20 s and extension at 72 °C for 20 s.
6. Then, the first PCR product is usually diluted 1:10 in double-distilled sterile water and used for the second PCR reaction, which can be performed in a 10-μl reaction volume containing 20 mM Tris (pH 8.3), 50 mM KCl, 1.5 mM $MgCl_2$, 200 μM of each deoxynucleotide triphosphate (dATP, dCTP, dGTP, dTTP), 0.5 μM each of forward and reverse primer and 0.25 units of Taq polymerase.
7. The methods of amplification visualization include gel electrophoresis using radioactive isotopes or silver staining, and more recently, automated sequencing equipment using fluorescent labeling.

11.5.1.5 Single-Strand Conformation Polymorphism (Protocol from Wistuba)

The basic methodological considerations for SSCP analysis from formalin-fixed, paraffin-embedded tissue are the following:

1. Intron primers flanking the exonic sequence of interest are frequently designed to bracket relatively short DNA fragments (smaller than 250 bp), thus allowing the detection of mutations at intron–exon junctions. If an exonic sequence is relatively large, two or more sets of primers need to be designed.
2. Nested or two-round (using the same set of primers in two consecutive amplifications) PCR methodology may be used. Moreover, a multiplex PCR protocol may be designed for the first round amplification.
3. The methods for amplification visualization include radioactive isotopes or gel electrophoresis silver staining.
4. SSCP analysis is done by heat-denaturing PCR products at 95 °C for 5 min followed by quick chilling on ice, and electrophoresis in nondenaturating polyacrylamide gels using several conditions, ideally in duplicate (i.e., room temperature/cold room, and with/without 10% glycerol).
5. Sample electrophoresis mobility should be compared with PCR products from control samples having wild and mutant sequences. Moreover, DNA extracted from normal tissues from the same individual tested is a good control for electrophoresis band mobility analysis.
6. For abnormal SSCP band confirmation of mutation, sequencing must be performed. Briefly, the abnormal PCR-SSCP product is excised from the gel, eluted, re-amplified by PCR and purified by running in an agarose gel. After gel extraction procedure, the obtained DNA is subject to sequence analysis using manual or automated sequencing.

11.5.1.6 Restriction Fragment Length Polymorphism (Protocol from Wistuba)

The most important methodological issues to consider when applying this methodology to formalin-fixed, paraffin-embedded tissues are:

1. PCR product availability has to be checked before restriction enzyme incubation.
2. The PCR product must be purified or cleaned before the use of the restriction enzyme in order to allow the correct enzymatic activity.
3. Negative and positive control digestions using reference-amplified wild-type and mutant DNAs is strongly recommended to monitor the efficacy of the restriction enzyme digestion.

   NOTE: Reproducibility of RFLP experiments needs to be addressed in addition to confirmation of mutated samples by sequencing.

11.5.1.7 Gene Methylation Analysis (Protocol from Wistuba)

For methylation-specific PCR (MSP) analysis using formalin-fixed, paraffin-embedded tissue samples, several modifications to the basic protocols have been suggested (Lee et al. 2002; House et al. 2003 from references Chapter 8):

1. One μg of tissue DNA should be treated with sodium bisulfite according to the appropriate protocols previously described (Lee et al. 2002; House et al. 2003 from references Chapter 8) to convert unmethylated cytosine to uracil.
2. The bisulfite-treated DNA needs to be column-purified over any DNA clean-up resin-based system, and then ethanol precipitated.
3. A nested two-step PCR approach in order to increase the sensitivity of detecting allelic hypermethylation at targeted sequences. It is highly recommendable to use a Taq polymerase system that includes a "hot-start" PCR approach.
4. A multiplex PCR protocol can be also designed to maximize the FFPET-extracted DNA. If multiplex PCR is used, step one of the nested MSP could be carried out with primer sets (sense and anti-sense) for up to four individual genes in each reaction. Step one primers flanked the CpG-rich promoter regions of the respective targeted genes. Hence, these primers did not discriminate between methylated and unmethylated nucleotides following bisulfite treatment.
5. PCR products of step one are diluted 1:1000 and subjected to the second step of MSP that incorporated one set of primers for each gene [labeled as unmethylated (U) or methylated (M)] designed to recognize bisulfite-induced modifications of unmethylated cytosines.
6. DNA isolated from normal peripheral lymphocytes from healthy individuals may serve as a negative methylation control. Human placental DNA treated in vitro with *Sss*I methyltransferase to create completely methylated DNA at all CpG-rich regions may serve as positive methylation control.
7. Products are analyzed on polyacrylamide or agarose gel electrophoresis.

11.5.1.8 Genotyping Assay of the D310 Repeat (Protocol from Wistuba)

1. At least 200 microdissected nuclei are used to amplify the D310 repeat from paired normal stromal cells and tumor/epithelial samples.
2. The primer sequences are 5′ACAATTGAATGTCTGCACAGCCACTT-3′ (sense) 5′GGCAGATGTGTTTAAGTGCTG-3′ (antisense).
3. A two-round PCR strategy may be utilized to amplify the D310 repeat from FFPETs. The PCR conditions used are 35 cycles using 60 °C as annealing temperature.
4. The methods for amplification visualization include gel electrophoresis utilizing radioactive isotopes or silver staining visualization.

11.5.1.9 In Situ Hybridization Protocol for Cytomegalovirus (Protocol from González)

1. Melt paraffin in the oven for 90 min at 60–70 °C.
2. Dewax sections in xylene 5 min, three times with slow shaking.
3. Rehydrate sections in ethanol 100% 1×, 95% 1×, 80% 1×.
4. Rinse in distilled water 2 min.
5. Pepsin digestion 10 min at 37 °C.
6. Pepsin solution: 70 mg pepsin in 70 ml distilled water preheated to 37 °C.
7. Post-fix immediately in paraformaldehyde 4%, 5 min at room temperature.
8. Wash in PBS-Triton, 3 min.
9. Incubate in glycine 10 mg/ml in PBS, 5 min, 37 °C.
10. Rinse in distilled water 1 min.
11. Dehydrate sections in ethanol 80% 1×, 95% 2×, 100% 2×.
12. Allow to dry at room temperature.
13. Incubate slides with CMV biotin probe diluted in hybridization mix. Hybridization mix: 45% formamide, 5×SSC, 25 mM/L sodium phosphate pH 6.5, 1×Denhardt's solution, 250 mg/ml salmon sperm DNA, 10% dextran sulfate, distilled water, 1 mg/ml CMV probe.
14. Denature DNA by placing sections in a waterbath at 92 °C±2 °C for 10 min.
15. Hybridization: may proceed at room temperature for 3 h or longer. Good results are obtained with 2 h at 37 °C followed by hybridization at room temperature overnight.
16. Remove coverslips by rinsing in 2×SSC.
17. Post-hybridization washings:

2×SSC	+0.1% SDS	2×	3 min
0.1×SSC	+0.1% SDS	2×	3 min
2×SSC	+0.1% SDS	1×	1 min

18. Block with 3% BSA in PBS-Triton, 10 min at 37 °C. Add 300–500 μl to each section.
19. Rinse with PBS and dry.
20. Incubation with SA-AP conjugate (streptavidine-alkaline phosphatase).
21. Rinse with PBS.
22. Wash slides with PBS-Triton solution, 3 min, three times with slow shaking.
23. Wash slides with PBS, 3 min, three times with slow shaking.
24. Wash slides with AP 9.6 solution, 3 min, three times with slow shaking.
25. Develop in NBT/BCIP solution.
26. Mount with Kayser's glycerol jelly.

11.5.2 RNA Analysis

11.5.2.1 Gene-Specific RT-PCR

11.5.2.1.1 Reverse Transcription

Reagents

1. DEPC-dH_2O (Research Genetics)
2. RNase inhibitor, 20 U/μl (Perkin Elmer)
3. dNTP, 250 μM (GenHunter)
4. First-strand buffer (Life Technologies)
5. Random hexamer primers, 50 mM (Perkin Elmer)
6. Superscript II (SSII) reverse transcriptase (Invitrogen)

Method

The protocol below may be used for amplifying individual transcripts from RNA recovered from microdissected cell populations as described in RNA-Based Studies of Microdissected Tissues.

TIP: The number of cells needed to perform gene-specific RT-PCR from microdissected samples is highly dependent on the quality of the tissue sample and the abundance level of the transcript(s) of interest. One thousand cells are recommended as a good starting point for gene-specific RT-PCR studies. However, reliable amplification can frequently be achieved using substantially fewer cells.

1. To RNA pellet, add 10 μl DEPC-H_2O and 1 μl RNase inhibitor.
2. Resuspend the RNA pellet with gentle tapping.
3. Quick spin.
4. Aliquot 5 μl into 2 sterile tubes for (+) and (–) RT reactions.
5. For each batch of samples, prepare additional control tubes as follows, using either high-quality RNA or DEPC-dH_2O in place of the 5-μl sample of RNA:

Control type	(+) RT	(–) RT
Positive	High-quality RNA	High-quality RNA
Negative	DEPC-dH_2O	DEPC-dH_2O

6. Prepare a sufficient volume of the following ±RT master reaction mixtures for all reaction tubes.

(+) RT master reaction mixture/tube	
1.0 μl	DEPC-dH_2O
2.0 μl	First-strand RT buffer
1.0 μl	dNTP
0.5 μl	Random hexamer primers
Total volume = 4.5 μl	

(–) RT master reaction mixture/tube	
1.5 μl	DEPC-dH_2O
2.0 μl	First-strand RT buffer
1.0 μl	dNTP
0.5 μl	Random hexamer primers
Total volume=5.0 μ	

7. Aliquot either 4.5 or 5.0 μl of the relevant mastermix to the (+) and (–) RT tubes.
8. Incubate at 65 °C for 5 min, then at 25 °C for 10 min.
9. Add 0.5 μl SSII to all (+) RT tubes only.
10. Incubate all tubes at 25 °C for 10 min, then at 37 °C for 40 min.
11. Incubate at 95 °C for 5 min to denature the SSII.
12. Quick spin.

11.5.2.1.2 Polymerase Chain Reaction (PCR)

Reagents

1. Advantage 2 Polymerase Mix, containing Taq polymerase and PCR buffer (Clontech).
2. dNTP, 25 μM (GenHunter).
3. Forward primer, 10 μM.
4. Reverse primer, 10 μM.
5. dCTP a-33P (or a-32P) (10 mCi/ml).
6. DEPC-dH_2O (Research Genetics).
7. High density TBE sample buffer, 5× (Novex).

Method

TIP: As with any PCR reaction, it is good practice to run duplicate or triplicate samples to ensure the validity of the PCR result.

TIP: Components and cycling will depend on individual template and primers.

1. Aliquot 3 μl of each cDNA sample into a sterile PCR tube.
2. Prepare sufficient volume of PCR master reaction mixture for all reaction tubes and add 7 μl to each tube.

PCR master reaction mixture/tube	
1.0 μl	PCR Buffer
0.8 μl	dNTP
0.2 μl	Forward primer
0.2 μl	Reverse primer
0.2 μl	dCTP a-^{33}P (or a-^{32}P) see TIP below
0.2 μl	Taq DNA polymerase, 5 U/μl
4.4 μl	DEPC-dH_2O
Total volume = 7 μl	

TIP: The PCR protocol presented here includes incorporation of radioactivity into the PCR products. Radioactivity is necessary for visualization on a denaturing acrylamide gel for low abundant transcripts or when PCR product patterns are complicated (i.e., polymorphic markers for LOH). The amount of radioactivity used in the above protocol often results in visible products in less than 2 h exposure. For abundant mRNAs, it may be possible to discern products on an ethidium bromide-treated agarose gel (replace ^{32}P or ^{33}P volume with water).

3. PCR Cycling Conditions:

Cycles	Temperature (°C)	Time
1	95	2 min
35	95	15 s
	X*	45 s
	72	5 min
1	72	10 min

X* is annealing temperature, dependent on the primer used.

4. Store the PCR products at 4 °C or continue to step 5.
5. Pour a 6% polyacrylamide sequencing gel while the PCR is cycling.
6. After cycling is complete, add 2.5 μl sample buffer (5×) to samples.
7. Denature samples at 95 °C for 3 min and place directly on ice.
8. Load 3.5-μl sample on gel and run at 1600 V to desired distance.
9. Dry gel and expose to phosphoimager screen or film as described under LOH protocol.

11.5.2.2 Targeted Differential Display

Materials

1. Stratagene RAP-PCR kit.
2. 5× RT buffer (GenHunter).
3. dNTP, 250 μM (GenHunter).
4. dNTP, 25 μM (GenHunter).
5. RNase inhibitor (Perkin Elmer).
6. Random hexamer primers (Perkin Elmer).
7. MMLV reverse transcriptase, 100 units/μl (GenHunter).
8. 10× PCR buffer (Perkin Elmer).
9. AmpliTaq DNA polymerase (Perkin Elmer).
10. DEPC-H_2O.
11. 3 M Na acetate, pH 5.2 (Life Technologies).
12. 100% ethanol.
13. 2× loading dye (95% formamide, 20 mM EDTA, 0.05% bromophenol blue, 0.05% xylene cyanol).

Methods

Important: Under sterile, RNase-free conditions

NOTE: The protocol was utilized to screen for tumor-specific alterations in transcript levels using a variety of screening approaches and primer combinations, including the RAP-PCR arbitrary primers from Stratagene and degenerate primers directed against known protein motif DNA sequences. Exact conditions must be determined for each differential display study depending on the number of cells utilized, the primer sets, and the number of transcripts the investigator wishes to screen.

11.5.2.2.1 Microdissection and RNA Isolation

1. Use frozen sections cut to a thickness of 12 μm.
2. Dissect approximately 5000–10,000 cells.
3. Complete the microdissection within 30 min of preparation of the tissue sections.
4. Isolate total RNA, as from protocol in RNA-based studies of microdissected tissues.

   TIP: Only use cells from frozen tissue that is extremely well-preserved.

   TIP: If anchored, oligo-dT-based differential display is to be performed (e.g., with the GenHunter kit), it is recommended to start with RNA recovered from a substantially larger number of microdissected cells.

11.5.2.2.2 Reverse Transcription

1. Add 24 μl H_2O and 1 μl RNase inhibitor, 20 U/μl, to RNA pellet.
2. Resuspend the RNA pellet with gentle tapping.
3. Quick spin.
4. Aliquot 12 μl into 2 tubes for the (+) and (–) RT reactions.
5. Prepare additional control tubes using the following in place of the sample RNA:
   a) A positive control prepared with known, high-quality RNA.
   b) A negative control prepared from either a LCM cap containing no tissue or RNase-free water.
   c) Prepare a sufficient volume of the following reaction mixture for each tube:

Reaction mixture/tube	
4 μl	5× RT buffer
2 μl	dNTP mix (250 μM)
1 μl	Random hexamer primers
Total volume = 7 μl	

6. Add reaction mixture to each tube in steps 3 and 4.
7. Incubate 10 min at 25 °C.

8. Add 1 μl MMLV reverse transcriptase to the (+) RT tube and the positive and negative controls.
9. Add 1 μl water to the (–) RT tube.
10. Incubate 10 min at 25 °C.
11. Heat 40 min at 37 °C.
12. Heat 5 min at 95 °C.
13. Store the cDNA at –20 °C until use.

11.5.2.2.3 PCR

TIP: Investigators must be especially careful when using this methodology to analyze archival tissue specimens. Formalin fixation in particular results in DNA that is difficult to amplify and often produces inconsistent PCR results, including artifactual allelic loss and poor amplification of large products. If this technique is to be utilized for analysis of archival samples, we highly recommend that replicate experiments (multiple independent dissections, triplicate PCR reactions, etc.) be used to verify results. Prepare sufficient volume of reaction mixture for all tubes:

Reaction mixture/tube	
1.0 μl	10×PCR Buffer
0.8 μl	dNTP mix (25 μM)
0.2 μl	Forward primer, 25 μM (see TIP 1)
0.2 μl	Reverse primer 25 μM
0.2 μl	^{33}P or ^{32}P, 20 mCi/ml,
0.2 μl	AmpliTaq DNA polymerase, 5 U/μl
6.9 μl	dH_2O
Total volume = 8.5 μl	

1. Add 1.5 μl cDNA (or DNA) from step 13 to each tube.
2. Add reaction mixture to each tube.
3. PCR cycling conditions in a PE 9600 thermocycler (see TIP 2)

Cycles	Temperature (°C)	Time (min)
1	94	2
35	94	0.5
	50	0.5
	72	2
1	94	2

TIP 1: In the published work cited above (Chuaqui et al. 1997: Urology Aug;50(2):302–307), degenerate primers directed against zinc finger motifs were used in conjunction with Stratagene's RAP-PCR arbitrary primers. To generalize this protocol, the PCR primers are written as forward primer and reverse primer. Please refer to the Stratagene protocol for more details.

TIP 2: The times and temperatures given above are samples only and what was used in the above referenced work. Please refer to the RAP-PCR protocol for details regarding the PCR conditions. We tried annealing temperatures varying from 50–58 °C.

11.5.2.2.5 PAGE

1. While the PCR is cycling, pour a 6% polyacrylamide sequencing gel.
2. After cycling is completed, add 10 µl 2× loading dye to samples.
3. Denature samples at 95 °C for 3 min, then place directly on ice.
4. Load 3.5-µl sample on gel.
5. Run at 55 W until the xylene cyanol migrates to the bottom of the gel.
6. Transfer gel to 3 mm Whatman paper and dry.
7. Expose to film in cassette with intensifying screen.
8. Develop film.
9. Look for differentially expressed bands between samples.

11.5.2.2.5 Sequencing of Differentially Expressed Bands

1. Excise the appropriate bands from the gel using a razor blade.
2. Place the piece of gel/paper in a microfuge tube.
3. Add 100 µl water.
4. After 2–5 min, peel the acrylamide off the paper.
5. Place the acrylamide in a new tube.
6. Grind the acrylamide with a pipette tip.
7. Add 100 µl water.
8. Heat at 94 °C for 10 min.
9. Precipitate the DNA using 0.1×volume 3 M sodium acetate and 2.5× volume 100% ethanol.
10. Resuspend the DNA pellet in 40 µl water.
11. To 12.2 µl DNA from step 10 above, add the following re-amplification reagents:

Reaction mixture/tube	
2.0 µl	10×PCR buffer
1.6 µl	dNTP 25 µM
2.0 µl	Forward primer, 25 µM
2.0 µl	Reverse primer, 25 µM
0.2 µl	AmpliTaq DNA polymerase
Total volume = 7.8 µl	

12. Use the same cycling conditions as in the first PCR, step 2 above.
13. Run out the PCR product on a 1.5% low melting point agarose gel.
14. Cut out the PCR product (approx. 100 µl), aliquot, and use directly in the sequencing reaction.

15. Sequence according to manufacturer's instructions. In the paper referenced below, we used the Perkin Elmer Amplicycle Sequencing kit.
    a) Chuaqui et al. (1997) Identification of a novel transcript up-regulated in a clinically aggressive prostate carcinoma. J Urol 50(2):302–307.

11.5.2.3 Nonisotopic in Situ Hybridization Protocol for mRNA Detection (Protocol from Gonzalez)

1. Heat slide at 75 °C for 10 min.
2. Wash in xylene, 5 min, twice.
3. Wash in methanol, 5 min, twice.
4. Wash in diethyl pyrocarbonate-treated water (DEPC-w), twice.
5. Digest formalin-fixed sections in 0.1% HCl-pepsin at 37 °C for 20 min; fresh samples are prefixed in paraformaldehyde and then incubated in proteinase K 1 mg/ml in PBS at 37 °C for 15 min.
6. Wash in DEPC-w, twice.
7. Postfixation in paraformaldehyde 4% (w/v) in PBS for 5 min at room temperature.
8. Wash in glycine 0.2% (w/v) in PBS for 5 min.
9. Digest control with RNAase A or DNAase I, for 30 min at 37 °C.
10. Wash in PBS, DEPC-w and finally methanol.
11. Dry at 37 °C.
12. Add hybridization cocktail.
13. Heat slides at 95 °C for 15 min.
14. Hybridization (15–60 min and/or overnight).
15. Wash in 2×SSC, 5 min, at room temperature.
16. With RNA probes, post-hybridization digest with RNAase A at 37 °C for 30 min.
17. Wash twice in 2×SSC, 5 min, at room temperature.
18. Block 10 min in TBT.
19. Incubate for 30 min in humid chamber with anti-digoxin monoclonal antibody 1:10,000 in TBS.
20. Wash in TBS.
21. Incubate with biotinilated anti-mouse IgG 1:200 in TBS.
22. Wash in TBS.
23. Incubate for 20 min in SA-AP 1:50 in TBT.
24. Wash in TBS.
25. Incubate with NBT/BCIP for 60 min.
26. Wash in tap water 5 min.
27. Dry at 37–75 °C, contrast with methyl green and mount in glycerol.
28. Negative controls.
    a) Omitted probe in hybridization solution.
    b) Irrelevant probes.
    c) Predigest with ribonuclease (RNase).

29. Positive controls.
    a) Probes for constitutive genes like actin.
    b) Known positive tissue sample.

11.5.2.4 Protocol for Transcript Amplification (Protocol from Watson)

NOTE: The protocol presented below was initially used by us to amplify RNA for GeneChip microarray analysis (Luzzi et al. 2001 from references Chapter 9). It is similar to a protocol more recently provided by Affymetrix http://www.affymetrix.com/support/technical/technotes/smallv2_technote.pdf) and is an alternative for those investigators who do not wish to utilize one of the many commercially available reagent systems referenced above. Enzymes and molecular reagents are purchased from the vendors indicated in parentheses.

1. Practice care in preventing RNase contamination. Generally, utilizing RNase-free reagents and disposables on a clean laboratory bench surface will suffice. DEPC treatment, high temperature baking, and RNase cleansing agents are usually not necessary.
2. Whenever possible, target amplification of all experimental samples should be performed in parallel. Utilize 'master mixes' for all reagents to avoid pipetting small volumes and to ensure uniformity of reagent concentrations across all samples. For *N* number of samples, make sufficient master mix for *N*+1 reactions.
3. Make certain that all components of master mixes are mixed thoroughly after addition of each component. Unless specifically instructed, do not vortex samples. Mix by repeatedly pipetting reagents in the tube, or gently mix the tube by hand. After mixing, briefly spin reaction tubes to collect the reaction mix at the bottom of the tube.
4. Reactions should be performed in a regulated heat block. A thermal cycler is ideal, particularly for 16 °C reactions. For prolonged incubations at 37 or 42 °C, it is important to occasionally (i.e., every 30–60 min.) mix the reaction components and then briefly spin the tube to collect any condensate. Prolonged incubations in a thermal cycler with a heated lid may increase the reaction temperature above 37 °C and therefore, should be avoided.

11.5.2.4.1 cDNA Synthesis and Purification (Round 1)

1. Add the following components to a 0.5-ml reaction tube:

10–100 ng RNA in nuclease-free water	11 µl
100 µM $T7T_{24}$ primer (Affymetrix)	1 µl
5′-GGCCAGTGAATTGTAATACGACTCACTATAGGGAGGCGGT$_{24}$-3′	

2. Mix the reagents, incubate at 70 °C for 10 min, spin briefly and place on ice.

3. Add the following components to the reaction:

5× Superscript II reaction buffer (InVitrogen)	4 μl
0.1 M DTT (InVitrogen)	2 μl
10 mM dNTP mix (InVitrogen)	1 μl

4. Mix reagents and incubate at 42 °C for 2 min. Then add 1 μl Superscript II RT: 100 U/μl (InVitrogen)
5. Incubate the final 20-μl reaction at 42 °C for 1 h.
6. Briefly spin reactions and place on ice.
7. Add the following components to the first-strand reaction for second-strand synthesis:

Nuclease-free water	91 μl
5× Second strand Buffer (InVitrogen)	30 μl
10 mM dNTPs (InVitrogen)	3 μl
10 U/μl DNA Ligase (InVitrogen)	1 μl
10 U/μl E. Coli DNA Polymerase I (InVitrogen)	4 μl
2 U/μl RNase H (InVitrogen)	1 μl

8. Gently tap tube to mix; spin briefly and incubate the 150-μl reaction at 16 °C for 2 h.
9. Add 2 μl (10 U) T4 DNA Polymerase (InVitrogen) to the reaction and incubate for 5 min at 16 °C.
10. Add 10 μl 0.5 M EDTA to stop the reaction. At this point, the double-stranded cDNA may be stored indefinitely at –20 °C.
11. Transfer the cDNA reaction to a 1.5-ml tube. Add 162 μl of phenol: chloroform: isoamyl alcohol solution to the second-strand cDNA reaction mixture, vortex for 30 s and then spin for 5 min at room temperature and 12,000×*g* to extract the cDNA.
12. Remove the aqueous (upper phase) into a clean 1.5-ml tube. Add 1 μl of 10 mg/ml glycogen and mix thoroughly. Add 0.5 volumes (~80 μl) of 7.5 M ammonium acetate and mix thoroughly. Add 2.5 volumes (~400 μl) of 100% ethanol and mix thoroughly to precipitate the extracted cDNA.
13. Centrifuge immediately at 16,000×*g* for 20 min at room temperature. Many downstream enzymes (e.g., T7 RNA polymerase) are inhibited by ammonium ions. Therefore, it is crucial that no salt is carried over during the precipitation step. Centrifugation should be performed at room temperature to avoid salt precipitation and all precipitated pellets should be thoroughly washed with 70% ethanol.
14. Wash the pellet thoroughly with 70% ethanol. Spin at 16,000×*g* for 5 min. Carefully pour off the 70% ethanol without displacing the pellet, although a pellet may not always be visible. Wash the pellet again, spin, and pour off the second wash. Carefully aspirate any remaining ethanol with a pipette, again being careful not to displace any visible pellet.

15. Allow the cDNA pellet to dry and dissolve it in 8 μl of nuclease-free water.

11.5.2.4.2 In Vitro Transcription and Purification (Round 1)

1. To synthesize complementary RNA (cRNA) from the double-stranded cDNA template, the Ambion Megascript in vitro transcription kit (T7 enzyme) is utilized. Following the recommended reaction conditions, add reaction components to the 1.5-tube containing the cDNA from step 15 above as follows:

cDNA from step 15 above	8 μl
10× Reaction buffer	2 μl
ATP Solution	2 μl
CTP Solution	2 μl
GTP Solution	2 μl
UTP Solution	2 μl
T7 Enzyme mix	2 μl

   Mix thoroughly and then transfer the entire reaction to a 0.5-ml tube. Incubate at 37 °C for 6 h. Gently mix the reaction by hand every hour and centrifuge briefly to collect reaction tube condensate. After in vitro transcription, the reaction may be frozen at –70 °C until purification.
2. cRNA is purified using Qiagen's RNeasy or Arcturus' PicoPure RNA purification columns. For Qiagen RNeasy columns, the manufacturer's recommended protocol is utilized. The cRNA is then eluted twice, with 40 μl of nuclease-free water.
3. Add 1 μl of 10 mg/ml glycogen, 0.5 volumes (40 μl) of 7.5 M ammonium acetate, and 2.5 volumes (200 μl) of 100% ethanol to the eluted 80 μl of cRNA. Mix thoroughly and spin immediately.
4. Centrifuge at 16,000×*g* for 30 min.
5. Wash the pellet, which may not be visible, thoroughly with 70% ethanol and allow the cRNA pellet to dry. Resuspend in 11 μl of nuclease-free water.
6. Depending on the initial amount of starting material, 1 μl of the synthesized cRNA may be quantified by UV absorbance or qualitatively assessed using the Agilent Bioanalyzer 2100.

11.5.2.4.3 cDNA Synthesis and Purification (Round 2)

1. Add the following components to a 0.5-ml reaction tube:

cRNA in nuclease-free water	11 μl
1 μg/μl Random hexamers	1 μl

   Mix the reagents, incubate at 70 °C for 10 min, spin briefly and place on ice.

2. Add the following components to the reaction:

5× Superscript II reaction buffer	4 μl
0.1 M DTT	2 μl
10 mM dNTP mix	1 μl

Mix reagents and incubate at 42 °C for 2 min. Then add 1 μl Superscript II reverse transcriptase (100 U/μl). Incubate the final 20-μl reaction at 42 °C for 1 h.
3. Briefly spin reactions and place on ice. Then add 1 μl (2 Units/μl) RNase H to the reaction and incubate for 20 min at 37 °C. Terminate the reaction by heating it to 95 °C for 5 min followed by a brief spin and incubation on ice.
4. To prime second-strand synthesis add 1 μl (100 μM) $T7T_{24}$ oligonucleotide primer to the reaction and heat at 70 °C for 10 min. Briefly spin the reaction and keep on ice.
5. Add the following second-strand synthesis reaction reagents:

Nuclease-free water	91 μl
5× Second-strand reaction buffer	30 μl
10 mM dNTP	3 μl
10 U/μl E. coli DNA Polymerase	4 μl

Incubate the 150-μl reaction at 16 °C for 2 h.
6. Add 2 μl (10 U/μl) T4 DNA Polymerase and incubate for 10 min at 16 °C.
7. Add 10 μl 0.5 M EDTA to terminate the reaction. The double-stranded cDNA may be stored at –20 °C until purification.
8. The cDNA is purified exactly as described in Section 11.5.2.4.1 steps 11–15. For subsequent generation of biotinylated aRNA targets for GeneChip microarray hybridization, the purified cDNA should be resuspended in 22 μl.

11.5.2.4.4 Biotin-Labeled in Vitro Transcription and Purification (Round 2)
NOTE: This final step should only be performed for microarray platforms that utilize biotinylated aRNA targets. The cDNA template created at the end of Sect. 1 may also be used as a template for in vitro transcription using any other direct or indirect labeling protocol. Alternatively, unlabeled aRNA may be generated again (as described in Sect. 11.4.2.2) which can then be used for cDNA synthesis with direct or indirect labeling.
1. The Enzo BioArray HighYield in vitro transcription kit is used to generate biotin-labeled cRNA for hybridization to Affymetrix GeneChip microarrays. The recommended protocol is utilized by mixing the following reaction components:

cDNA template	22 μl
10× HY reaction buffer	4 μl
10× Biotin label ribonucleotides	4 μl
10× DTT	4 μl
10× RNase inhibitor mix	4 μl
20× T7 RNA polymerase	2 μl

Add the reaction components to the 1.5-ml tube containing the cDNA from step h above. Mix thoroughly and then transfer the entire reaction to a 0.5-ml tube. The reaction is carried out for 6 h at 37 °C, mixing the reaction components every hour and spinning briefly to collect the reaction condensate.

2. After 6 h, the reaction may be frozen at –70 °C until purification
3. The biotin-labeled cRNA is purified using Qiagen's RNeasy or Arcturus' PicoPure RNA purification columns. For Qiagen RNeasy columns, the manufacturer's recommended protocol is utilized. The cRNA is eluted with 40 μl of nuclease-free water and the eluate is passed over the column a second time to increase the yield.
4. Quantify 1 μl of the purified, labeled cRNA by UV absorbance. The cRNA may also be qualitatively assessed by gel electrophoresis or Agilent Bioanalyzer RNA Nano Chip assay. The amplified cRNA should have a molecular weight distribution between 500 and 1500 nucleotides. A very low or very high molecular weight cRNA distribution suggests anomalous amplification and will generally result in a poor hybridization signal if applied to the microarray.
5. Fragment and hybridize labeled cRNA using standard protocols supplied by Affymetrix or other relevant microarray manufacturer.

11.5.2.5 Protocol for RT-PCR Validation (Protocol from Watson)

NOTE: The protocol detailed below is one that we routinely use to validate gene expression patterns of transcripts identified by microarray studies. It utilizes SybrGreen dye (in the form of a commercially available enzyme mix) and standard PCR primers. The approach works equally well for RNA derived from both frozen and FFPE LCM tissue.

11.5.2.5.1 Primer Design

1. Full-length reference mRNA sequences for target genes of interest should be collected. The NCBI's Locus Link site (http://www.ncbi.nlm.nih.gov/LocusLink/) is a useful source for these sequences.
2. Using each mRNA sequence, primers may be designed using a number of programs. Applied Biosystem's Primer Express software is often supplied with thermocycler instrumentation. If this software is not available, Primer3 (http://www-genome.wi.mit.edu/cgi-bin/primer/prim-

er3_www.cgi) is an excellent web-based program for primer design. GeneRunner (http://www.generunner.com/) is also a freely available nucleic acid analysis software package that includes primer design functionality.

3. When designing primers, the following guidelines are recommended
   - The amplicon size should range between 70 and 150 nucleotides. If primers are being used with RNA derived from FFPE tissue the amplicon size should be less than 100 nucleotides, if possible.
   - Each primer should have a theoretical melting temperature of 58–60 °C. Primers are generally 20–23 nucleotides in length and range from 45–65% G/C content.
   - At the 3′ end of each primer, no more (and preferably less) than 2 of the last 5 bases should be C or G. If possible, the 3″ nucleotide should be an A or T.
   - To exclude amplification of possible contaminating genomic DNA, at least one primer should span across two exons.
   - If cDNA synthesis is to be primed only with oligo-dT, primers should be located within 1000 nucleotides of the polyA tail to ensure efficient first-strand cDNA synthesis of the amplification target sequence.

11.5.2.5.2 cDNA Synthesis

NOTE: For cDNA synthesis with 20–50 ng of RNA from LCM tissue, we utilize Qiagen's Sensiscript reverse transcriptase enzyme. The protocol is the same for RNA derived from FFPE or frozen tissue.

1. Create an enzyme "master mix" for enough samples + 10% (e.g. if 10 samples are run, make enough "master mix" for 11 samples). Per reaction:

10× RT buffer	2 μl
5 mM dNTP mix	2 μl
10 μM random hexamer	2 μl
10 μM dT_{21} primer	2 μl
RNase inhibitor (10 U/μl)	1 μl
Sensiscript RT enzyme	1 μl

   Mix thoroughly by hand, spin briefly, and aliquot 10 μl to the required number of reaction tubes. If a large number of reactions are planned, cDNA synthesis may be conveniently performed in a 96 well thermal cycler plate.

2. Add 20–50 ng of RNA (the same amount of RNA for each sample) in a volume of 10 μl of nuclease-free water to each reaction tube or well. Mix thoroughly by gently pipetting up and down several times.
3. Incubate at 37 °C for 60 min.

4. Freeze reactions at –20 °C until use in PCR. It is important to avoid multiple freeze-thaws of the cDNA as this will compromise cDNA stability and amplification performance.

11.5.2.5.3 Primer Validation

NOTE: Prior to utilization with valuable LCM RNA samples, it is important to establish that primer pairs will specifically amplify their intended target sequence in a linear and reproducible fashion. To accomplish this, serial dilutions of a cDNA control cocktail are amplified, C_T values calculated, and amplicons assessed by both melt curve analysis and/or gel electrophoresis.

1. Convert 1 µg of control RNA into cDNA by priming with random hexamers (1 µM) and oligo dT (1 µM) using standard cDNA synthesis protocols. A useful source of "control" RNA can be derived from a single 50-µm serial section from the same tissue block utilized for LCM. If primers are to be used with RNA derived from LCM of FFPE tissue, then control RNA should be similarly derived.
2. Add 80 µl of nuclease-free water to the 20 µl of cDNA reaction product to yield a final concentration of 10 ng input RNA/µl. Make the following set of serial dilutions, sufficient to validate 10 primer pairs

Tube 1:	40 µl of cDNA (10 ng/µl)
Tube 2:	20 µl cDNA + 20 µl nuclease-free water (5 ng/µl)
Tube 3:	10 µl cDNA + 30 µl nuclease-free water (2.5 ng/µl)
Tube 4:	5 µl cDNA + 35 µl nuclease-free water (1.2 ng/µl)

3. For each primer pair to validate, set up a primer/enzyme master mix for 10 reactions (4 cDNA dilutions in duplicate + 2 no template controls)

ABI 2×SybrGreen master mix	10 µl
10 µM Forward primer	2 µl
10 µM Reverse primer	2 µl
Nuclease-free water	76 µl

4. Aliquot 18 µl of mix across ten wells in a single row of a 96-well optical PCR plate. Each row of the plate will contain a different primer set for validation.
5. To two adjacent wells of each plate (e.g. A1, A2) add 2 µl of each dilution of cDNA. The last two wells of the row (e.g. A9, A10) will serve as no template controls.
6. After plate setup, reactions should be run using the standard quantitative PCR thermal profile appropriate for the instrumentation. After completion of the run, a melt curve profile of the amplicons should be performed. In addition, amplified products may be analyzed by 2.5% agarose gel electrophoresis or Agilent Bioanalyzer 2100 and DNA Chip assay to verify a uniform amplification product is present of the expected size.

11.5.2.5.4 Quantitative PCR

NOTE: After establishing that primer sets specifically amplify their intended target sequence with suitable linearity, cDNA generated from LCM tissue may be assayed. In general, each PCR assay should be performed in duplicate or triplicate to provide satisfactory precision. Depending upon transcript abundance and primer amplification efficiency, 2.5–5 ng equivalents of input RNA from frozen LCM tissue should be sufficient to generate amplification signals below a C_T of 35 cycles. This can be readily inferred with data generated from the standard curve as described above. Accordingly, 50 ng of RNA should be sufficient to validate the expression pattern of as many as ten transcripts. When working with RNA derived from FFPE tissue, as much as 10–20 ng equivalents of input RNA may be necessary to achieve similar C_T values.

1. For each primer set, create an enzyme "master mix" for: [(number of samples + 1 no template control) ×2, (duplicate reactions)]+10% (e.g. if 10 samples are run, make enough "master mix" for 24 samples). Per reaction:

ABI 2× SybrGreen Master Mix	10 µl
10 µM Forward primer	0.2 µl
10 µM Reverse primer	0.2 µl
Nuclease-free water	7.6 µl

2. Aliquot 18 µl of each mix into the wells a 96-well optical PCR plate.
3. Add 2 µl of each sample cDNA (Sect. 11.5.2.5.2, cDNA Synthesis) to duplicate wells of the plate.
4. Reactions should be run using the standard quantitative PCR thermal profile appropriate for the instrumentation.

11.5.3 Protein Analysis

11.5.3.1 2-D Polyacrylamide Gel Electrophoresis

Solutions

TIP: Use electrophoresis grade reagents to prepare the following solutions:

1. IEF lysis buffer (50 ml)
   a) Add 21 g urea to 35 ml HPLC-grade H_2O in a 50-ml Falcon tube (final concentration 7 M).
   b) Vortex vigorously for several minutes.
   c) Add:

7.6 g	Thiourea
2 g	Chaps
0.5 g	Mega 10
0.5 g	OBG
250 μl	Triton X-100
0.25 g	Tris
0.4 g	DTT
500 μl	Pharmalytes or IPG buffer pH 3–10 (Amersham)
500 μl	β-Mercaptoethanol

   d) Add 10 μl tributylphosphine 2 mM, under the hood (2 mM final concentration).
   e) Add bromophenol blue as indicator.
   f) Check volume is 50 ml.
   g) Vortex until all is dissolved (or attach tube to a rotator).
   h) Aliquot 1 ml into microfuge tubes.
   i) Store at –20 °C.
2. 10× Electrophoresis running buffer (0.25 M Tris, 1.92 M glycine, 1 M SDS) – 10 l
   a) Add 300 g Tris-base, 1441 g glycine, and 10 g SDS to ~7 L HPLC-grade H_2O.
   b) Mix gently until dissolved.
   c) Bring volume to 10 l.
3. 30% Acrylamide stock (1 l)
   a) Add 292 g acrylamide and 8 g piperazine diacrylamide (PDA) to 700 ml HPLC-grade H_2O, under the hood.
   b) Stir to dissolve.
   c) Bring volume to 1 L.
   d) Filter through 0.45-μm-pore size filter.
   e) Store at 4 °C in the dark.
4. Separating acrylamide gel
   a) Below are the solution volumes required to prepare one 9–18% gradient gel. Prepare sufficient volume for the number of gels to be run.

Solution	Volume units	9% Gel	18% Gel
1.5 M Tris-HCl, pH 8.8	ml	11.5	11.5
20% SDS	ml	0.23	0.23
30% Acrylamide	ml	14	28
TEMED	μl	11.7	11.7
10% APS	μl	117	117
HPLC-grade H_2O	ml	20	6
Total	ml	45.8	45.8

5. Equilibration buffer I (50 ml)
   a) Mix together 18 g urea and 10 ml of 0.5 M Tris-HCl, pH 6.9.

   b) Vortex vigorously.
   c) Add 10 ml of 20% SDS and 200 mg DTT.
   d) Invert gently several times.
   e) Add 15 ml glycerol.
   f) Vortex vigorously (or attach tube to a rotator for 10–15 min).
   g) Add bromophenol blue as indicator.
6. Equilibration buffer II (50 ml)
   a) Same as equilibration buffer I EXCEPT add 5.0 g iodoacetamide INSTEAD of DTT.
7. Transfer buffer (1 l)
   a) 25 mM Tris-HCl (3 g)
   b) 190 mM glycine (14.44 g)
   c) 20% methanol (200 ml)

Methods

Day One: Sample Preparation

1. Preparation of sample (see processing of microdissected tissue for molecular analysis-protein-based studies section).
2. Reswelling
   a) Remove Immobiline Drystrips (Amersham Pharmacia Biotech) from –20 °C and allow to equilibrate at RT.
      TIP: The pH range of the strip used should be the same as the pH range of Pharmalytes or IPG buffer used in the IEF lysis buffer.
   b) Notch the basic end of the strip to mark each sample.
   c) Load the first sample into Reswelling trays (Immobiline DryStrip Reswelling Tray/Pharmacia #18–1004–31.).
   d) Place DryStrips gel-side down into each slot.
   e) Remove air bubbles by pressing down with a pipette tip.
   f) Overlay completely with DryStrip Cover fluid (Amersham Pharmacia, #17–1335–01).
   g) Repeat for every sample including the MW standard (2-D SDS-PAGE standards, pH range 4.5–8.5, MW 17,500–76,000, Bio-Rad, #161–0320).
   h) If samples are concentrated in one region of the strip, redistribute by pipetting.
   i) Cover the tray with the lid.
   j) Incubate overnight at RT to allow the strips absorb the samples.

Day Two: First Dimension

1. Clean the electrophoresis chamber (Pharmacia LKB Multiphor II) and the Immobiline strip tray and wipe out with paper towels and Kimwipes to remove mineral oil.
2. Place the tray on top of 50 ml DryStrip Cover fluid.
3. Remove strips.
4. Place on Whatman paper gel-side up.

TIP: Placing the strips gel-side down might result in protein loss and gel damage.

5. Leave for ~1 min.
6. Place into the electrophoresis chamber gel-side up.
7. Arrange the strips so that their edges are in one line.
   TIP: Time is important to prevent crystallization.
8. Wet the pre-made IEF electrode strips (Amersham Pharmacia, #18–1004–40) with HPLC-grade H_2O.
9. Dry slightly between two pieces of Whatman paper.
10. Place two buffer strips on both edges of the strips and perpendicular to them, covering the top of the bromophenol blue on each side.
11. Make sure that the square end of each strip is at the cathode (black/–) end and the pointed end is at the anode (red/+) end and also that the anode and cathode electrode ridges are in the correct orientation.
12. Overlay liberally with DryStrip Cover fluid between the Immobiline strips and outside the electrodes.
13. Electrophorese for 36–48 h, using the following sequence of settings:

	Voltage (V)	Amps (mA)	Wattage (W)	Time (h)
1	500	100	33 W	0.05
2	500	110	70 W	1
3	3500	141	32 W	5
4	3500	70	38 W	Until stopped

14. The bromophenol blue should be seen migrating towards the anode within at least 1 h. By next day, the strips should be colorless.
15. By the next day, if the bromophenol blue has not disappeared, running can be paused and the electrode strips can be replaced from whichever side. Continue running until the dye has disappeared.

Day Three: Second Dimension

1. Prepare apparatus.
   a) Wash and scrub plates very well in soap and hot water.
   b) Rinse in ddH_2O.
   c) Leave the plates to air-dry or wipe with methanol-soaked Kimwipes.
   d) Order plates in Protean-II Multi-Gel casting Chamber (Bio-Rad, #165–2025) as follows:

Bottom of chamber
Small plate, 20 cm
Spacers, 20 cm×1 mm
Eared plates, 20 cm
Spacers, 20 cm×1 mm
Large plate, 20 cm
Meylar sheet
Repeat as needed

   e) Fill the chamber with acrylic blocks.
   f) Tighten the screws.
   g) Tape the edges of the chamber to prevent leakage.
2. Prepare gradient acrylamide gel (9–18%)
   a) Add 9% gel solution to the center compartment of the distributor (mixing chamber) and 18% gel to the peripheral compartment (reservoir chamber).
   b) Start the magnetic stirrer in the mixing chamber.
   c) Remove air bubbles from the tubing by opening the valve slowly.
   d) Allow the tubing to fill with gel solution, then close the valve.
   e) Turn on the stirrer.
   TIP: The stirring bar should be between the openings between the mixing and reservoir compartments, but not on top of them.
   f) Open the valve between the mixing and the reservoir chambers (upward).
   g) Make sure the 18% solution is flowing into the mixing compartment.
   h) Open the valve to start the flow of the acrylamide solution into the Protean chamber.
   i) Allow reasonable flow. Fast flow results in loss of a gradient, whereas a slow flow results in polymerization of the solution in the tubing). In addition, the rate of flow changes with time due to the change of pressure. However, the chamber should be filled in at least 10 min.
   j) Stop when the gel has reached 0.5 cm from the top of the glass plates.
   k) Overlay the gels carefully with HPLC-grade H_2O using a syringe.
   l) Cover with Saran Wrap.
   m) Allow to polymerize overnight.

Day Four: Second Dimension (continued)
1. Remove strips.
2. Place on Whatman paper, gel-side up, for 1 min.
3. Equilibrate in equilibration buffer I for 10–15 min.
4. Set up gels in electrophoresis chamber (Protean II Multi-cell, Bio-Rad).
5. Make sure there is no leakage.
6. Equilibrate strips in equilibration buffer II.
7. Remove strips.

8. Place on Whatman paper one by one, gel-side up.
9. Identify notches.
10. Cut ~1 in. from both sides.
11. Place gels with basic side closer to the anode (red/+) and the acidic side closer to the cathode (black/–) to be consistent.
12. Run gels at 40 mA/gel for 15 min or until dye front is about 1 inch into the gel.
13. Turn down voltage and run at 65–70 V constant overnight (If in a hurry, set power supply at 360 mA and 500 V and run for 8 h).

Day Five: Transfer and Sequencing of Proteins

1. Transfer onto PVDF or nitrocellulose membranes.
   a) Soak gel in transfer buffer for 5–10 min.
   b) Soak pre-cut transfer membrane and Whatman paper in transfer buffer.
   c) Wet the transfer apparatus with transfer buffer.
   d) Place 3–4 Whatman papers on bottom of the transfer apparatus.
   e) Place membrane on top and remove air bubbles.
   f) Place gel on top of the membrane and remove air bubbles.
   g) Place 3–4 Whatman papers on top of the gel and remove any air bubbles.
   h) Transfer for 3 h at no more than 25 V (1–1.5 mA/cm^2 of the gel/membrane). Transfer for 5 h if more than 5–6 gels.
2. Gel staining.
   a) See staining protocols.
3. Excision of Protein Spots (for sequencing by MS)
   TIP: Work under the hood to prevent the gel from being exposed to too much air, thus avoiding contamination with keratins.
   a) Xerox the gel and assign the spot(s) to be sequenced.
   b) Cut out the protein spot for sequencing with a pipette tip.
   c) Remove to a microfuge tube.
   d) Chop up the gel piece with pipette tip.
   e) Add a solution of 50% methanol/10% acetic acid/40% H_2O to the gel pieces.
   f) Incubate for 30 min.
   g) Spin down.
   h) Remove the supernatant.
   i) The sample is ready for mass spectrometry sequencing.

11.5.3.2 Immunoblot

Materials

1. Protease inhibitor cocktail (Boehringer Mannheim)
2. 2× SDS sample buffer (Novex)
3. Pre-casted gels (e.g., Novex)

4. Molecular weight protein standard (e.g., MultiMark Multi-Colored Standard, Novex)
5. 25× Tris-glycine transfer buffer (e.g., Novex)
6. 100% Methanol
7. Blocking solution containing casein (e.g., Pierce)
8. Wash solution I (for 1 l: 1 g ovalbumin (Sigma), 1 g dry milk, 1 ml Tween-20 (Bio-Rad), 100 ml 10× PBS, 900 ml distilled water).
9. Primary antibody
10. Horseradish peroxidase-conjugated secondary antibody
11. Wash solution II (for 1 l: 1 ml Tween-20, 100 ml 10× PBS, 900 ml distilled water)
12. Luminol and oxidizing reagents (e.g., Pierce or NEN)

Method

1. Protein separation by polyacrylamide gel electrophoresis.
   a) On top of the transfer film (containing the microdissected cells), continuously pipette up and down 10–20 µl 2× SDS sample buffer for 2–3 min in order for the proteins to go into solution.
   b) Transfer protein solution to a screw-top tube and boil the sample(s) for 3 min.
   c) Spin briefly to collect any condensed fluid.
   d) Set up pre-casted gel and 1× Tris-glycine-SDS running buffer.
   e) Load all of the sample buffer containing the protein onto the gel). Include a molecular weight protein standard.
   f) Perform electrophoresis at 110 V for 1.5–2 h.
2. Electrophoretic transfer to a membrane (nylon, pvdf or nitrocellulose).
   a) Wash gel in 1× transfer buffer (Novex) for 30 min.
      TIP: Some sources do not recommend this step as smaller proteins may diffuse out of the gel.
   b) Set up the electrophoretic transfer. Soak 4–5 sponges and 2 pieces of filter paper in 1× transfer buffer. If using PVDF membrane, briefly soak the membrane in 100% methanol, then 1× transfer buffer.
   c) Assemble apparatus from negative electrode to positive electrode as follows: 2 sponges, filter paper, gel, membrane, filter paper, 2–3 sponges.
   d) Perform transfer at 40 V for 2 h.
3. Primary and secondary antibody incubations
   a) Remove membrane and place in a blocking solution containing casein for 30 min at RT.
   b) Wash 3×10 min in wash solution I.
   c) Incubate with primary antibody overnight at optimal titer (e.g., 1:1000 in wash solution I) while rocking at 4 °C.
   d) Wash 3×10 min in wash solution I.
   e) Incubate with a horseradish peroxidase-conjugated secondary antibody for 1 h at optimal titer (e.g., 1:20,000 in wash solution II) while rocking at 4 °C.

f) Wash 3×10 min in wash solution II.
TIP: Alternatively, the secondary antibody may be biotinylated where a streptavidin-peroxidase conjugate binds to the biotin on the secondary antibody, greatly increasing the sensitivity for protein detection.

4. Visualization
   a) For the chemiluminescent reaction, incubate with equal parts of luminol and oxidizing reagents (Pierce or NEN, Boston, MA) for 5 min at RT.
   b) Develop by X-OMAT autoradiography per recommendation of the manufacturer.

11.5.3.3 Immunohistochemistry

11.5.3.3.1 Protocols from Pathogenetics Unit

Important: All tissue sections should be cut onto slides which have been charged, silanized, or poly-lysine coated in order to prevent detachment of tissue during the procedure.

1. Protocol for frozen tissue sections.
   a) Fix the tissue with 70% ethanol for 15 min
   b) Wash with PBS for 5 min
   c) Block endogenous peroxide with Proxoblock for 45 s (Zymed Labs)
   d) Wash with PBS for 5 min
   e) Mark the tissue borders using a pap-pen.
   f) Apply blocking antibody (reagent A) according to kit directions for 10 min (Zymed Histostain-plus kit, 2nd generation lab-SA detection system kit, cat No: 85–9743, CA, Tel: 1–800–874–4494).
   g) First antibody for 1 h (or overnight at 4 °C, according to manufacturer's data sheet).
   h) Wash in PBS 3×, 5 min total.
   i) Secondary antibody (reagent B)from kit for 10 min when using primary mouse, 20 min when using primary rabbit.
   j) Wash in PBS 3×, 5 min total.
   k) Apply enzyme conjugate (reagent C) for 10 min.
   l) Wash in PBS 3×, 5 min total.
   m) Color reaction (AEC or DEB; reagent D) for 5–10 min.
   n) Wash in ddH_2O water.
   o) Counter-stain with hematoxylin for 3 min.
   p) Wash in tap water.
   q) Apply Permount to the tissue section and coverslip
2. Protocol for ethanol-fixed-paraffin-embedded tissue sections.
   a) Put in 60 °C oven for 1 h.
   b) Xylene #1 for 7 min.
   c) Xylene #2 for 7 min.

d) Ethanol 100%, 5 min with shaking.
e) Ethanol 96%, 5 min with shaking.
f) Ethanol 70%, 3 min with shaking.
g) Tap water, 3 min.
h) ddH_2O, 3 min.
i) Put the slides for 20 min in a solution of 0.3% H_2O_2 (30%) in methanol absolute (to 300 ml methanol add 5 ml H_2O_2).
j) Wash for 5 min in PBS.
k) Wash for 5 min in ddH_2O.
l) Mark the tissue borders using a pap-pen.
m) Apply blocking antibody (reagent A) according to kit directions for 10 min (Zymed Histostain-plus kit, second generation lab-SA detection system kit, Cat. No: 85–9743, CA, Tel.: 1-800-8744494).
n) First antibody for 1 h.
o) Wash in PBS 3×, 5 min total.
p) Secondary antibody (reagent B) from kit for 10 min when using primary mouse, 20 min when using primary rabbit.
q) Wash in PBS 3×, 5 min total.
r) Apply enzyme conjugate (reagent C) for 10 min.
s) Wash in PBS 3×, 5 min total.
t) Color reaction. Shake the battle well before applying reagent D for 10 min.
u) Wash in ddH_2O water.
v) Counter-stain with hematoxylin for 3 min.
w) Wash in tap water.
x) Apply Permount to the tissue section and coverslip

3. Protocol for formalin-fixed and paraffin-embedded tissue sections.
   a) Prepare 3–5-μm sections. Note: better to use freshly cut sections or to keep the sections in −20 °C until use.
   b) Mark the slides with the correct date and the antibody + dilution to be used on the section (write only with a pencil on the slide).
   c) Put in 60 °C oven for 1 h.
   d) Xylene #1 for 5 min.
   e) Xylene #2 for 5 min.
   f) Ethanol 100%, 12 dips.
   g) Ethanol 96%, 12 dips.
   h) Ethanol 70%, 12 dips.
   i) Tap water 12 dips.
   j) ddH_2O, 12 dips.
   k) Special pretreatment may be required for sections according to manufacturer's instructions prior to binding of the primary antibody:

If **NO** pretreatment required:

a) Put the slides for 20 min in a solution of 0.3% H_2O_2 (30%) in absolute methanol
b) Wash for 5 min in PBS
c) Repeat step 2
d) Wash for 5 min in ddH_2O
e) Continue with step l below

Steam pretreatment:

a) Put the slides for 10 min in a solution of 3% H_2O_2 (30%) in absolute methanol
b) Wash for 5 min in ddH_2O
c) Repeat.
d) Heat-treat the slides according to:
   1. Place slide in a Coplin jar containing 10 mM sodium citrate buffer, pH 6.0
   2. Cover with a vented plastic wrap and place in steamer (Black and Decker) for 20 min
   3. Take out the jar and let the sections cool in the jar for 20 min at RT
e) Go to step l below

Microwave pretreatment

a) Place sections in a microwave-safe dish
b) Put 1500 ml 10 mM citrate buffer with 0.1% Tween 20 in a pressure cooker (total volume 1500 ml for 20 slides)
c) Put the dish inside the cooker and place in microwave for 40 min
d) Cool for 20 min
e) Wash for 5 min with PBS
f) Repeat
g) Go to step a below

Pronase pretreatment:

a) To 60 ml PBS at 37 °C add 0.03 gm pronase (Sigma-Aldrich, cat No: P6911)
b) Keep at 37 °C and stir
c) Put slides into solution (while leaving at 37 °C) for 10 min
d) Wash twice with PBS (5 min each time)
e) Put the slides for 10 min in a solution of 3% H_2O_2 (30%) in absolute methanol
f) Wash 2× in ddH_2O for 5 min
g) Go to step a below

Trypsin pretreatment:

a) Take out the trypsin (Sigma, 1 g; Cat. No: T-7409; Tel: 314–771–5750). Dilute to aliquots of 0.1%: 100 mg in 100 ml and store at –20 °C in quantities of 500 µl
b) Using a pipette, cover the tissue section with a 500-µl aliquot on the slide
c) Leave for 8 min in 37 °C (incubator not covered)
d) Wash twice with PBS (5 min each time)
e) Put the slides for 10 min in a solution of 3% H_2O_2 (30%) in absolute methanol
f) Wash 2× in ddH_2O for 5 min
g) Go to step b
l) Mark the tissue borders using a pap-pen
m) Apply blocking antibody (reagent A) according to kit instructions for 10 min (Zymed Histostain-plus kit, second generation lab-SA detection system kit, cat No: 85-9743, CA, Tel.: 1-800-8744494)
n) Apply primary antibody for 1 h (or overnight at 4 °C, according to manufacturer's data sheet)
o) Wash in PBS 3×, 5 min total
p) Apply secondary antibody (reagent B) from kit for 10 min when using primary mouse, 20 min when using primary rabbit
q) Wash in PBS 3×, 5 min total
r) Apply enzyme conjugate (reagent C) for 10 min
s) Wash in PBS 3×, 5 min total
t) Color reaction (AEC or DEB; reagent D) for 5–10 min
u) Wash in ddH_2O water
v) Counter-stain with hematoxylin for 3 min
w) Wash in tap water
x) Apply Permount to the tissue section and coverslip

11.5.3.3.2 Immunohistochemical Staining Protocol for p53 Using Streptavidin-Biotin-Complex Technique (Protocol from González)

1. Cut 4–5 µm paraffin sections and deparaffinize.
2. Block endogenous peroxidase activity with 3% H_2O_2 in methanol for 15 min.
3. Unmask p53 protein antigen by immersing sections in 0.3 M buffer citrate and heat to boiling in a microwave oven at high level three times 7 min each.
4. Incubate slides in normal rabbit serum for 20 min.
5. Apply mouse monoclonal antibody against p53 protein (clone DO-7; Dakopatts, California) diluted 1:50 for 90 min at room temperature.
6. Wash sections in TBS and apply the bridge-antibody consisting of biotinylated-anti-mouse IgG diluted 1:300 in buffer TBS for 30 min.
7. Wash in TBS and incubate in streptavidin-biotin-peroxidase complex diluted 1:100 for 30 min.
8. Wash in TBS and apply diamino-benzidine and 0.04% H_2O_2 in distilled water for 10 min.
9. Intensify the reaction by incubating in 0.1% OsO_4 in TBS for 3 min.
10. Slightly counterstain with nuclear-fast red and mount.

11.6 Legal Issues, Informed Consent

Before investigators study clinical samples, it is imperative that they protect the patient's privacy and obtain informed consent. Before a patient is included in any clinical research study, the physician must fully inform him or her of the treatment, its benefits, and possible side effects. The patient then signs a consent form which shows his or her understanding and acceptance of these issues. In molecular profiling studies, patients should also clearly understand how and why their tissues will be used, and be aware that their clinical information will be integrated into the analytical database.

The following give an in-depth discussion of patient informed consent and confidentiality:

1. Websites
   a) Ethical, legal, and social implications of human genetics research: (http://www.genome.gov/)
   b) Guide to understanding informed consent: http://www.cancer.gov/clinicaltrials/AGuidetoUnderstandingInformed-Consent/
   c) Protecting participants in clinical trials: http://www.cancer.gov/clinicaltrials/digestpage/protecting-participants
   d) Health Insurance Portability and Accountability (HIPAA) Resources: http://www.hhs.gov/ocr/hipaa/finalreg.html; http://privacyruleandresearch.nih.gov/; http://www.cap.org:80/apps/docs/hipaa/hipaa_articles-html
2. Articles
   a) Sobel ME (1999) Ethical issues in molecular pathology: paradigms in flux. Arch Pathol Lab Med 123(11):1076–1078
   b) Sobel ME, Wolman SR (1999) The ethical uses of tissues in research. Breast J May 5(3):153–155
   c) Sobel ME, Wolman SR (1999) Ethical considerations in use of human tissues in research. Cytometry 38(4):192–193
   d) Grizzle W, Grody WW, Noll WW et al. (1999) Recommended policies for the uses of human tissue in research, education, and quality control. Ad Hoc Committee on Stored Tissue, College of American Pathologists. Arch Pathol Lab Med 123(4):296–300
   e) Qualman SJ, France M, Grizzle WE, LiVolsi VA, Moskaluk CA, Ramirez NC, Washington MK (2004) Establishing a tumour bank: banking, informatics, and ethics. Br J Cancer 90:1115–1119

Subject Index

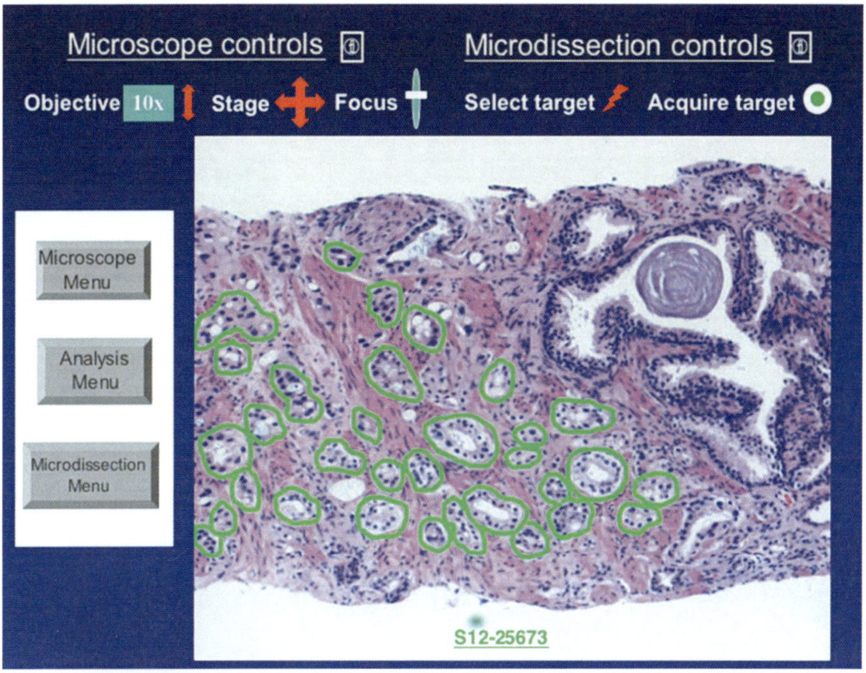

Microscope controls
Microdissection controls
Objective
10x
Stage
Focus
Select target
Acquire target
Microscope Menu
Analysis Menu
Microdissection Menu
S12-25673

Fig. 1.2

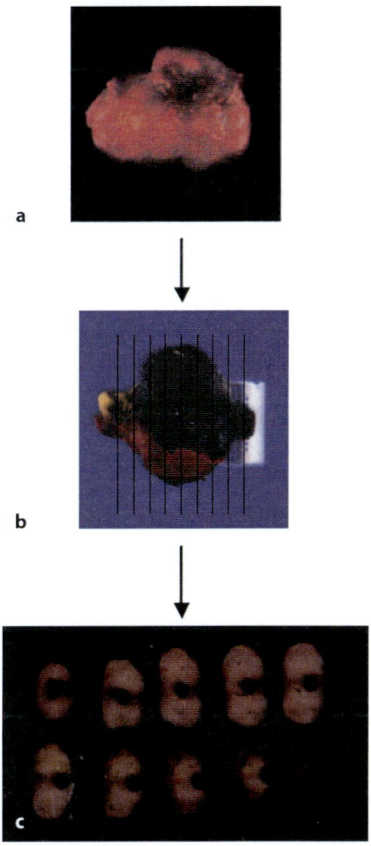

a
b
c

Fig. 3.2

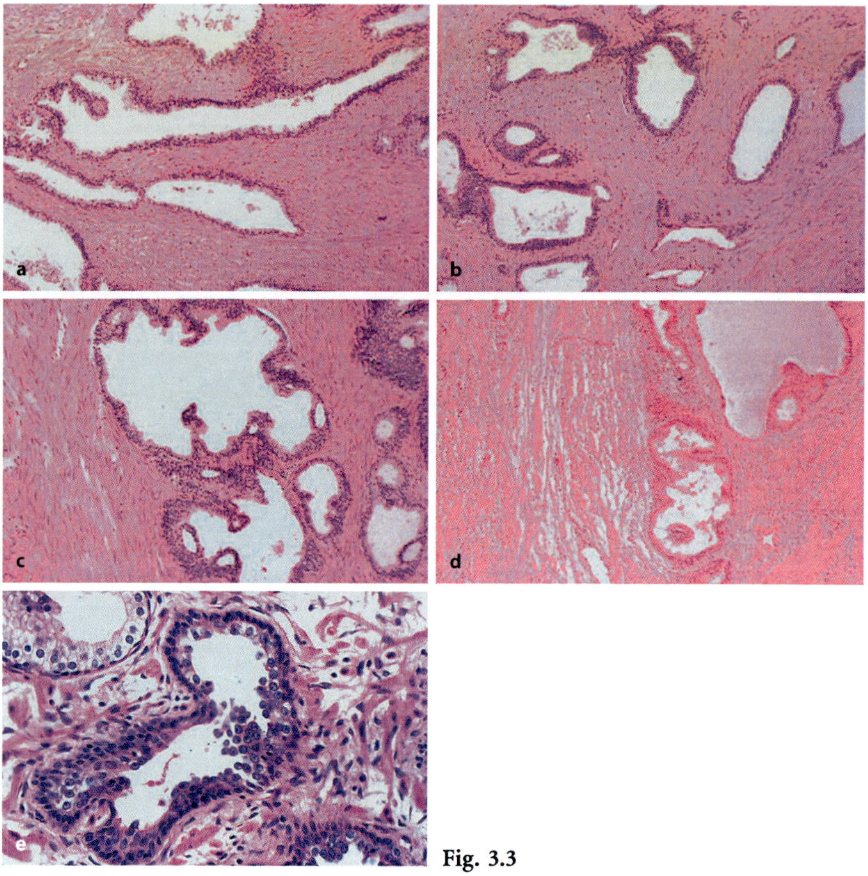

Fig. 3.3

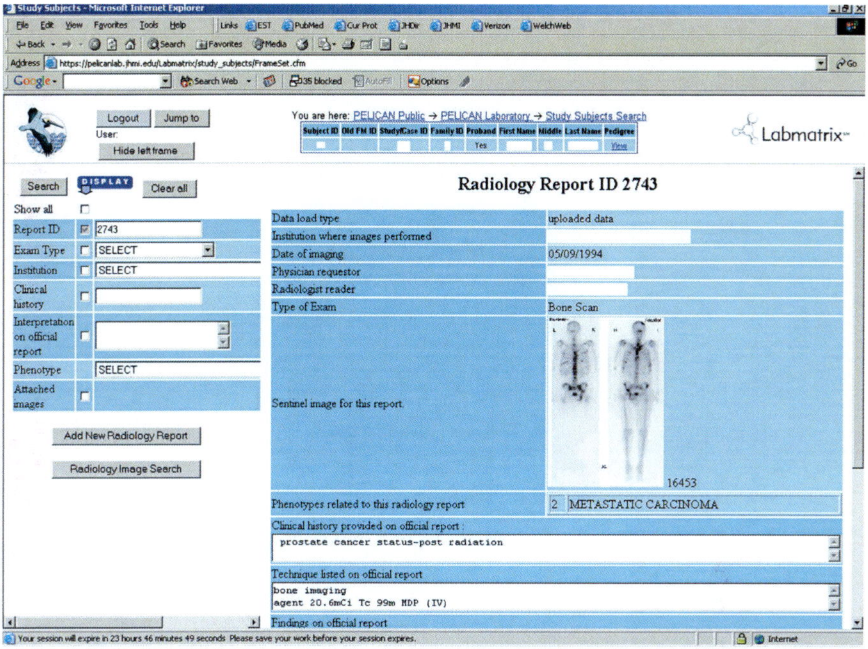
Study Subjects - Microsoft Internet Explorer
Address https://pelicanlab.jhmi.edu/Labmatrix/study_subjects/FrameSet.cfm
Logout
Jump to
User:
Hide left frame
You are here: PELICAN Public → PELICAN Laboratory → Study Subjects Search
Subject ID
Old FM ID
Study/Case ID
Family ID
Proband
First Name
Middle
Last Name
Pedigree
Yes
View
Labmatrix
Search
DISPLAY
Clear all
Show all
Report ID
2743
Exam Type
SELECT
Institution
SELECT
Clinical history
Interpretation on official report
Phenotype
SELECT
Attached images
Add New Radiology Report
Radiology Image Search
Radiology Report ID 2743
Data load type
uploaded data
Institution where images performed
Date of imaging
05/09/1994
Physician requestor
Radiologist reader
Type of Exam
Bone Scan
Sentinel image for this report
16453
Phenotypes related to this radiology report
2 METASTATIC CARCINOMA
Clinical history provided on official report :
prostate cancer status-post radiation
Technique listed on official report
bone imaging
agent 20.6mCi Tc 99m MDP (IV)
Findings on official report
Your session will expire in 23 hours 46 minutes 49 seconds. Please save your work before your session expires.
Internet

Fig. 4.1

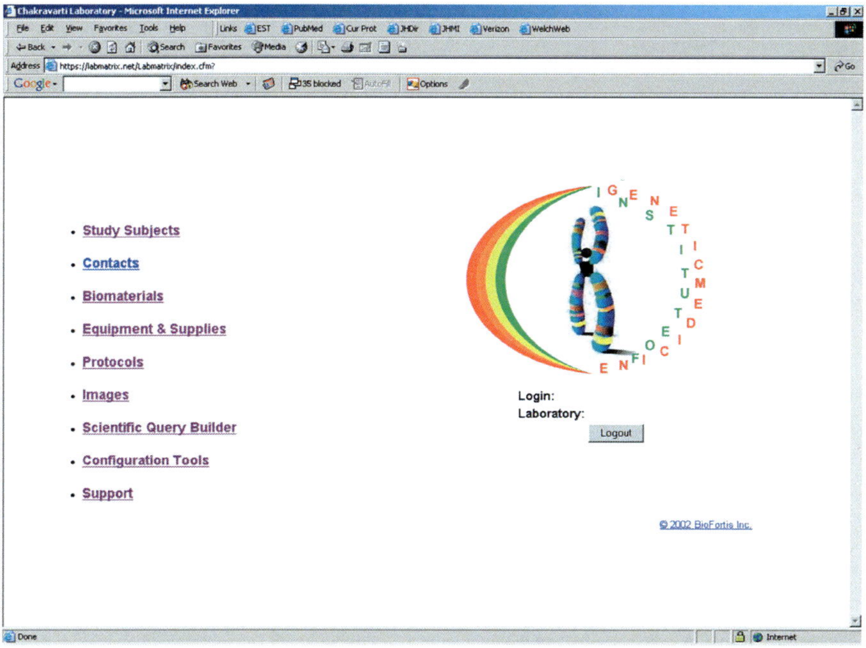
Chakravarti Laboratory - Microsoft Internet Explorer
Address https://labmatrix.net/Labmatrix/index.cfm?
Study Subjects
Contacts
Biomaterials
Equipment & Supplies
Protocols
Images
Scientific Query Builder
Configuration Tools
Support
Login:
Laboratory:
Logout
© 2002 BioFortis Inc.
Done
Internet

Fig. 4.2

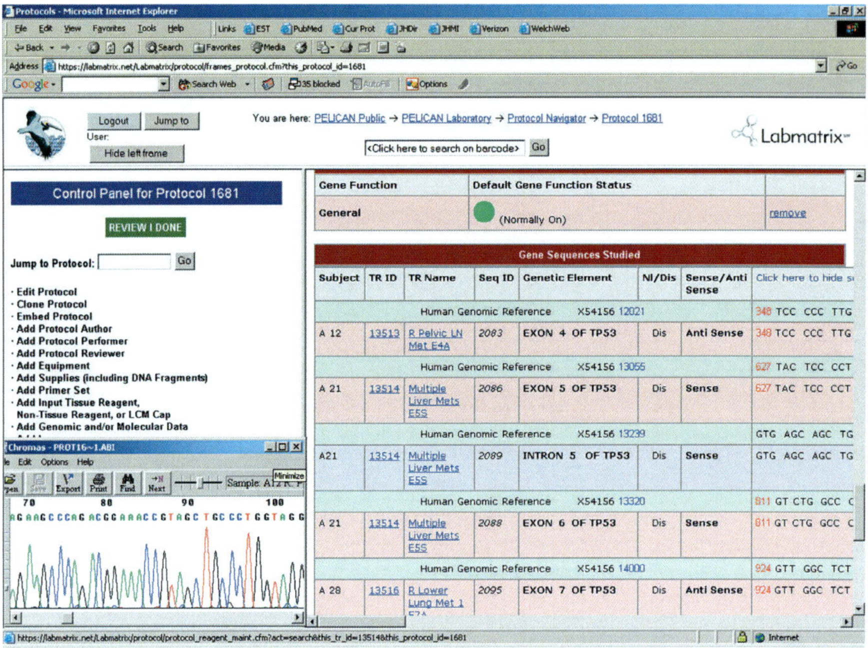

Fig. 4.3

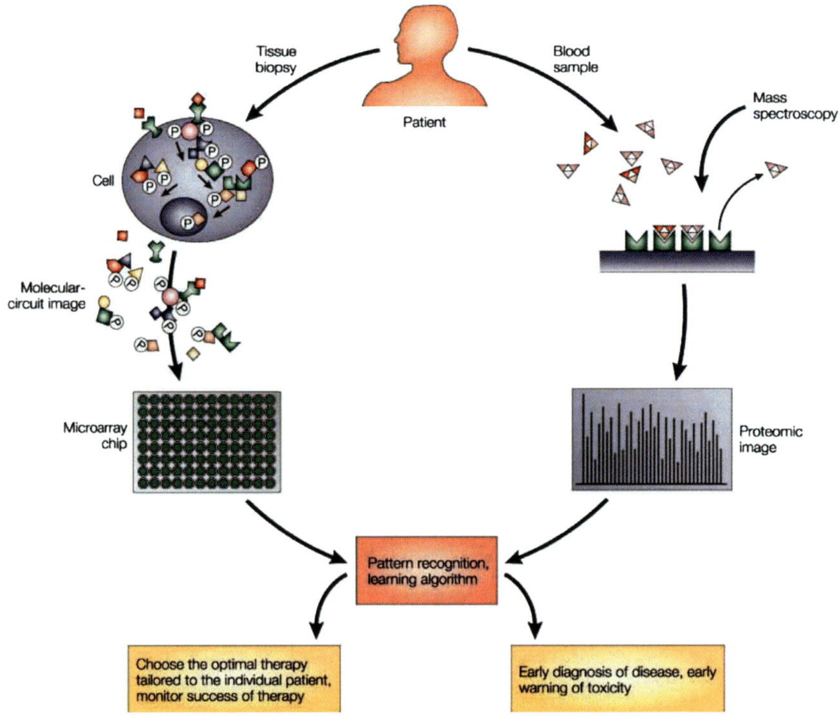

Fig. 10.1

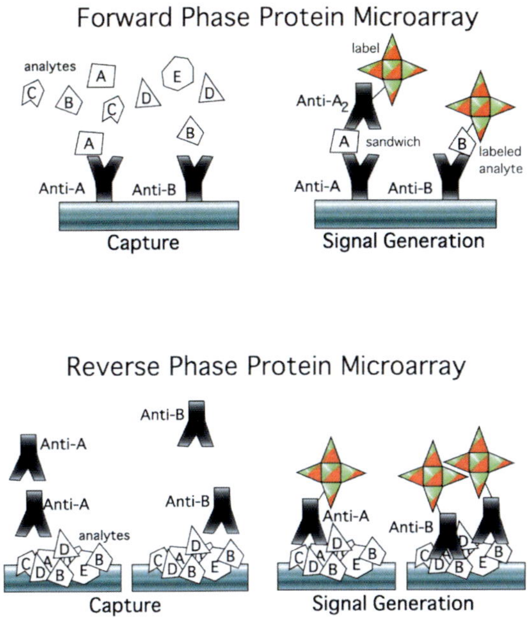

Fig. 10.2

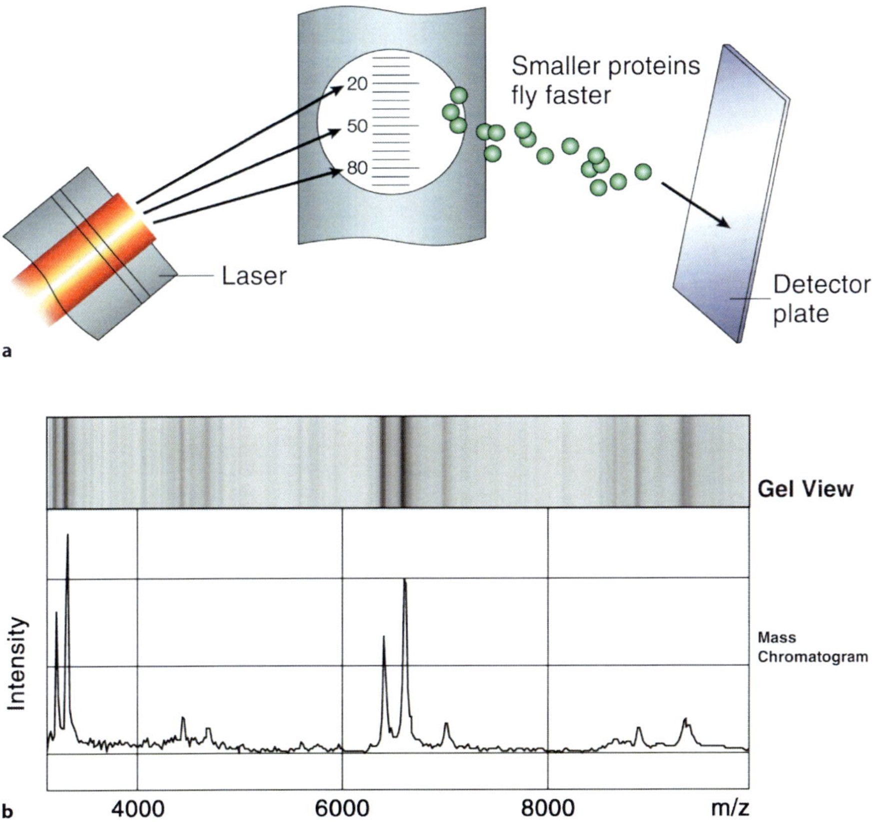

Fig. 10.3